T0211869

Transformations in Schooling

Transformations in Schooling

Historical and Comparative Perspectives

Edited by Kim Tolley

First published in 2007 by
PALGRAVE MACMILLAN™
175 Fifth Avenue, New York, N.Y. 10010 and
Houndmills, Basingstoke, Hampshire, England RG21 6XS.
Companies and representatives throughout the world.

PALGRAVE MACMILLAN is the global academic imprint of the Palgrave Macmillan
division of St. Martin's Press, LLC and of Palgrave Macmillan Ltd. Macmillan® is a registered
trademark in the United States, United Kingdom and other countries. Palgrave is a registered
trademark in the European Union and other countries.

ISBN 978-1-349-53464-7   ISBN 978-0-230-60346-2 (eBook)
DOI 10.1057/9780230603462

Library of Congress Cataloging-in-Publication Data is available from the Library of
Congress.

A catalogue record for this book is available from the British Library.

Design by Macmillan India Ltd.

First edition: April 2007

10 9 8 7 6 5 4 3 2 1

Contents

List of Tables

Acknowledgments

Many colleagues have contributed to this project. For their initial enthusiasm and suggestions, I would like to thank Anil Belvadi, Nancy Beadie, Craig Campbell, and Geoffrey Sherington. In 2004, several scholars joined this initial group in two panel presentations at the American Educational Research Association (AERA) annual meeting in San Diego, including Heather Kimberly Dial, Victoria-María MacDonald, and Reva Joshee. Lauri Johnson joined us for dinner and eventually participated in the project as a coauthor. Thanks to Nancy Beadie, Peter Kallaway joined this project. Craig Campbell and Geoffrey Sherington brought Tim Allender on board, and Victoria-María MacDonald invited Mark Nilles to coauthor a chapter. Conversations with Kay Whitehead at AERA and Meri Clark at the History of Education Society led to their participation in this project. Ting-Hong Wong also graciously offered to participate, and his chapter greatly benefits the book.

Many historians and sociologists have provided feedback on various chapters or on the overall project. Among these, I would like to thank Gary J. McCulloch, Blythe F. Hinitz, Harry Smaller, and Roberta Wollons for their thoughtful feedback.

For encouragement and support of this project, I thank Amanda Johnson, education editor at Palgrave Macmillan, and her assistant, Emily Leithauser. The anonymous external reviewers provided insightful and relevant feedback. I also thank John Sollami and Elizabeth Sabo for adroitly shepherding this volume through production, and the editorial staff at Macmillan India Ltd. for improving the final manuscript with their meticulous copyedits.

Introduction: Historical and Comparative Perspectives on Transformations in Schooling

Kim Tolley

D
o state-supported systems of schooling contribute to social inequalities, or do they disrupt them? Do community-based schools facilitate increased social mobility, or do they simply contribute to a growing divide in wealth and social status between the rich and the poor? What are the conditions under which educational systems change? Such questions have surfaced recently in current policy debates about the likely effect a free market would have on modern schooling. Critics of highly centralized, state-supported schools argue that education would respond to free market incentives with increased innovation, higher academic achievement, and stronger community relations, all of which would lead to increased social mobility among previously disenfranchised groups.[1] On the other hand, their opponents argue that free markets have never served as vehicles of equity, and that both publicly and privately funded schools would suffer from the effects of increased choice and competition.[2] These are perennial concerns. Since 1776, when Adam Smith discussed the "Education of Youth" in *The Wealth of Nations*,[3] policymakers have argued over the extent to which highly centralized educational systems either enhance or inhibit social inequalities.

Periods of early national schooling in different geographic areas of the world provide fascinating sites for the investigation of such questions. The transformation from colony to nation has often accompanied nascent efforts at school building or school reform. In many societies, such periods brought emerging state-supported systems of education into competition with market-based and church- or community-sponsored schools. Such periods bring into focus conflicts and collaborations between members of different social classes and ethnic groups. They also highlight tensions and outright conflicts between the state's effort to promote a national identity and the attempts of local communities to preserve unique and separate cultures. These struggles can express themselves through the forms of schooling supported by local communities as alternatives to the educational systems funded and supported by the state.

In simplest terms, this book explores two related questions: One, during periods of early national educational transformation, what factors have enabled various groups to renegotiate power in both state-supported and market-based systems of schooling? Two, how has the hegemony[4] of powerful classes or political groups renewed or reasserted itself during the early process of nation building, and how has it been, in the words of Raymond Williams, "resisted, limited, altered, and challenged by pressures not at all its own"?[5]

One of the questions that social historians have addressed recently is why systems of education developed at different rates and in different forms across countries. Even in countries with similar colonial origins, such as Canada, the United States, and Australia, institutional structures and policies have evolved in dissimilar ways. In the past several decades, several theories of the relation between education and state formation have emerged to account for such differences.

Margaret Scotford Archer formulated the first theoretical frame of reference for the development of educational systems. In *Social Origins of Educational Systems,* she presented a historical and structural comparison of the educational systems in France, Denmark, England, and Russia. Archer theorized that changes in educational systems arise as a result of group interactions that are conditioned or influenced, but not determined, by prior structural or social factors. For Archer, such interaction includes group conflict, the development of political alliances, and the elaboration of effective ideologies. "Change occurs because new educational goals are pursued by those who have power to modify previous practices." Archer theorized that two forms of challenge to the state are possible: *substitution,* a process whereby groups create rival institutions to ensure an educational provision compatible with their needs and wants, or *restriction,* a process whereby groups gain access to national legislative machinery. Archer argued that in all systems, education becomes increasingly integrated with the state over time, but in different ways. She distinguished between centralized and decentralized systems, arguing that systems with restrictive origins tended to become highly unified and systematized, whereas systems with substitutive origins tended to develop with weak forms of unification and strong forms of differentiation and specialization.[6]

In *Education and State Formation,* Andy Green argued that the development of public education systems could only be understood in relation to the process of *state formation,* "the historical process by which the modern state has been constructed." According to Green, because the intervention of the state affects the formation of national education systems, it is therefore the nature of the state that explains the particular national forms and timing of the development of school systems in different countries. After comparing national systems of schooling in England, France, and the United States, he concluded that the formation of national systems developed more swiftly in countries where the process of state formation was most intensive, either as a result of (1) military threats or territorial conflicts, (2) revolution or successful struggle for national independence, or (3) state-level motivation to embrace educational reform to escape from relative economic underdevelopment. Green concluded that centralized states created centralized educational bureaucracies, whereas more liberal states, such as the United States, created more decentralized systems. In all states, regardless of their degree of centralization, class relations determined the purposes of schooling, because "the different forms of hegemony operating between the dominant and subordinate classes... was ultimately responsible for what schools did, for who they allowed to go to what school and for what they taught them when they were there."[7]

Scholars agree that broad theories of the relationship between education and state formation must be tested against close-grained studies of schooling in local contexts. Margaret Archer has argued for a "continuous interplay between the theoretical and the comparative analysis of social structure," noting that the act of constructing social systems a priori and then fitting comparative data to the systems will never contribute much to our knowledge of social structures.[8] The historians of American education Carl Kaestle and Maris Vinovskis have criticized "one-to-one models of history" that have attempted to relate education to a single aspect of social change, including claims that factory production caused educational reform, urbanization promoted school bureaucracy, capitalism caused increased enrollments, or that modernization resulted in increased literacy. They have urged social historians to give more attention to issues of localism, to try to understand the difference in schooling patterns in rural and urban areas, and to explore relations among various communities in different geographical areas and within differing social groups.[9] In a similar vein, the critical theorist Michael Apple has urged scholars to "think contextually." Noting that the real relations of hegemony in society require close-grained empirical study on multiple levels, he recommended that researchers examine specific relations of power at each level and consider the ways that relations of economy may interact with culture.[10]

This book brings together a group of scholars with the aim of "thinking contextually" about specific periods of transformation in education history. The historians and sociologists whose work is collected here do not subscribe to one particular theoretical perspective or ideological belief: this is a diverse group. What binds them together is their critical appreciation of context and their quest to understand the complex interactions that have given rise to varying forms of schooling in different parts of the world.

In the following chapters, contributors from Australia, Canada, South Africa, Taiwan, and the United States consider a number of questions: What factors influenced the evolution of different forms of school governance and funding in some of the former colonies of the Spanish and British empires? In cases where schooling became accessible to women and ethnic minority groups, what factors contributed to increased access and participation? How can we interpret the transformations of varying kinds of educational practice in these different countries? Some authors investigate the interrelation of the state and local communities in the creation and support of systems of education. Others investigate the access and entry of women and ethnic minorities to schoolrooms and the teaching profession during different historical periods and in varying geographic areas. Several scholars analyze the way that various ethnic groups have struggled with the state to define their identity. Although their chapters draw from a number of theoretical perspectives, each addresses issues concerning the state, community, identity, and access to formal schooling.

The book is organized into four sections. The chapters in the first section, "Education and State Formation," pay close attention to the way that local economic, social, and political contexts shaped the emergence of early national systems of schooling in Australia, Singapore, India, and Colombia. The chapters in the second section, "Politics, Ideology, and Policy," explore the ideological and political context in which education policy evolved in Canada and South Africa. The chapters in the third section, "The Market, the State, and Transformations in Teaching," explore the gender shift in teaching that occurred during early national periods in Australia and the United States. The chapters in the final section, "Culture, Identity, and Schooling," investigate the means by which the transformation from colony to nation entailed reinterpretations of culture and social identity, transformations that affected access to schooling in different regions of the United States.

Education and State Formation

The transformation from colony to nation has often included school building or school reform. The chapters in this section investigate early national schooling by paying close attention to issues of localism and relations among various communities, classes, and political constituencies.

Middle-Class Formations and the Emergence of National Schooling: A Historiographical Review of the Australian Debate

Geoffrey Sherington and Craig Campbell review the Australian historiography on the origins of national schooling, focusing on class formation and schooling in the Australian colonies. As is often the case in the asking of new questions, the present school choice behaviors of the Australian middle class draws their attention to the nineteenth century.

Drawing on a rich array of nineteenth- and twentieth-century historical studies, Sherington and Campbell argue that the dominant interpretations of the emergence of "national" schooling in Australia have tended to neglect or deal unsympathetically with the emergence and continuing presence of the "private" school sector, whether those schools were private-venture academies and colleges, or whether they were church- or state-supported grammar schools. Yet these schools were the overwhelmingly dominant providers of anything that might be thought of as secondary or "higher" education in the nineteenth century. Very often their purposes and practices produced a powerful and competing vision of both "nation" and "national" education.

Education and State Formation Reconsidered: Chinese School Identity in Postwar Singapore

Ting-Hong Wong defines *state formation* as the "historical trajectory through which a governing regime builds or consolidates its dominance."[11] What distinguishes his theoretical approach from that of earlier scholars is his emphasis on the dialectical nature of this process. Wong argues that state formation is never a unitary project, because ruling authorities must deal with contradictory demands as they attempt to build national identity, win support from subordinated groups, and outmaneuver powerful political opponents. As a result, state educational policy can have unanticipated results.

Wong argues that some scholars have underestimated the relative autonomy of public school systems in some states and have failed to consider the extent to which a ruling group can establish its dominance by incorporating the cultures of subordinated groups. Because such acts of incorporation ultimately influence the nature of the state itself, Wong theorizes that state formation and education are related in an interactive, dialectical, and recursive manner.

State Schooling in the Raj: Disengagement and Resistance

Tim Allender's broad survey of nineteenth-century colonial education efforts in India examines the consequences of the imposition of systemic state-directed schooling. He postulates that as the century progressed, the increasingly active hand of the state in education contributed to its own disengagement from the broader Indian population. His chapter investigates the means by which greater state supervision over Indian education to ensure the teaching of English in most government-funded schools resulted in the marginalization of thousands of language and religious indigenous schools. Allender argues that state government in India became an unintentional agency for the stimulation of stimulating national resistance narratives and activist anti-British organizations such as the Arya Samaj.

Struggles for Schooling after the Independence Wars in Colombia, 1820–1830

Poverty and other factors contributed to the conflicts over the implementation of a national school system in Colombia. In the first decade of Colombian independence, the new government sought to implement a policy in which local towns financed their own schools while the state trained teachers in provincial capitals. The early national government insisted that local towns make every effort to ensure that boys and girls of every class and race could enroll in public schools. In this chapter, Meri L. Clark demonstrates that local communities resisted this policy for a number of reasons. By the 1830s, powerful private associations had emerged and had assumed responsibility for maintaining both public and private schools. Clark concludes that various pressures on state resources constrained the first decade of school formation, leading the state to shift many obligations to private hands.

The cases in Clark's chapter reveal the diffusion of the Enlightenment ideal that universal access to primary education could improve society. Nevertheless, local conflicts over which children could attend school, what they would study, and how their schools would be funded provoked the creation of powerful private associations. By the early 1830s, the government began to allow these associations to assume responsibility for maintaining public schools and establishing private ones.

Politics, Ideology, and Policy

Historic Equity and Diversity Policies in Canada

In their study of historic diversity and equity policies in Canada, Reva Joshee and Lauri Johnson argue that policies such as the Royal Proclamation and the Quebec Act laid the foundation for the development of a discursive framework that ultimately helped to create policies and programs that addressed diversity and equity in education. However, Canada's diversity policies were not simply a top-down creation of the state. Noting that most nongovernmental organizations have been ignored in previous histories of the development of Canadian policies for diversity in education, Joshee and Johnson demonstrate that government officials, community activists, and educators worked together to shape and reshape the web of policies surrounding and supporting diversity in education.

Although the role of most nongovernmental organizations has been ignored in accounts of the development of Canadian policies for diversity in education, archival evidence indicates that labor organizations worked alongside religious and ethnocultural groups and civil liberties organizations to protest educational segregation and exclusion, to introduce intercultural education programs, and to lobby for diversity and equity policies. Educational organizations worked with community groups and government agencies to conduct antidiscrimination seminars, produce and distribute curriculum materials, and organize conferences to explore issues of diversity. Joshee and Johnson argue that through these processes the policy actors helped to create an enduring public commitment to diversity that has not erased the long-standing systems of oppression but has the power to disrupt them.

The Development of a Conference and Policy Culture: The New Education Fellowship and British Colonial Education in Southern Africa

Peter Kallaway's chapter explores the political and ideological context from which British colonial education policy arose in the interwar era. In the context of the Depression and the rise of totalitarianism in Germany, Italy, Japan, and the USSR, Kallaway traces a clear shift in emphasis at the conferences of the New Education Fellowship (NEF), from Progressive-Era pedagogy of personal and individual development to a hardnosed appraisal of policies that promote economic growth and social development in a democratic context.

By analyzing changes in educational policy in Africa, Kallaway attempts to monitor the emergence of alternative voices at international conferences by the mid-1930s. His sources include documents related to the conferences of the NEF, British Colonial Office policy, the conferences of the International Missionary Council (IMC), U.S. foundations, and other significant networks of educational policy debate. Kallaway charts the place and role of South African participation in these events in the context of the rise of political opposition to imperialism following World War I and the establishment of the League of Nations.

Education Markets, the State, and Transformations in Teaching

Two chapters in this book consider issues of gender in the transformation of schooling within different geographical regions in Australia and the United States. They focus on the following questions: In cases where schooling became accessible to women, what factors contributed to increased access and participation? Both chapters emphasize the importance of local communities, economic contexts, education markets, and state policy in restricting or enhancing men's and women's access to various forms of schooling.

The Teaching Family, the State, and New Women in Nineteenth-Century South Australia

Kay Whitehead's chapter explicates the construction of teaching as gendered work in the context of changes in patriarchal relations in nineteenth-century South Australia. It explores the notion of the "teaching family" prior to state intervention in schooling, identifying men's, women's, and children's social and economic contributions to the family unit, and explains that the teaching family comprised husband-and-wife teaching teams, various combinations of parents and children, and all-female families.

Under the 1851 Education Act, the teaching family was co-opted by the state to accommodate the demand for sex-segregated schooling. Governing authorities upheld the patriarchal household by granting most licenses to male teachers as household heads and principal breadwinners in the family economy, thus protecting their positions. With the introduction of compulsory schooling in 1875, however, the state as employer began to employ teachers individually and differentiate their wages on the assumption that the men would marry and that the women would be single. In effect, the state substituted the teaching family with married men and single women, and marginalized married women.

Whitehead argues that although the reconstruction of teaching as waged labor shored up the patriarchal household by constructing men as sole breadwinners, women were not entirely disempowered as teachers. Indeed, she demonstrates that the individuation of wages facilitated the economic and social conditions for single women teachers, discursively positioned as "new women," to individually and collectively contest the established gender order by the end of the nineteenth century.

Transformations in Teaching: Toward a More Complex Model of Teacher Labor Markets in the United States, 1800–1850

From the eighteenth to the late nineteenth centuries, school teaching in the United States transformed from a predominantly male occupation to a predominantly female occupation. Nancy Beadie and I undertake a comparative study of this transformation, focusing on developments in New York, a northern state, and North Carolina, a southern state. Unlike other studies that have focused on the interactions between the state and local communities in an attempt to explain why education systems change over time, this chapter examines the shift from male to female teachers that occurred from 1800 to 1850, a period that preceded the expansion of large public school systems in the United States.

We investigate the role of the education marketplace in facilitating the entry of women to the occupation of teaching. We conclude that supply and demand in the education market played a role in facilitating the access of women to teaching positions during the early national period. Although most studies of teacher wages and the so-called feminization of teaching in the nineteenth century focus exclusively on state-funded teaching in common schools, Beadie and I demonstrate that teaching in academies and other voluntary schools was significant for structuring female participation in the occupation. By the time large state-funded education systems developed in the later nineteenth century, the question of whether women would teach in public schoolrooms had already been resolved. In many cases, the state appropriated the structures and processes developed in community-based, private, and voluntary schools. Our findings contribute to the possibility of developing a more nuanced model of teacher labor markets in the antebellum era.

Culture, Identity, and Schooling

To varying degrees, the transformation from colony to nation entailed reinter-pretations of culture and social identity. In this book, the term *colonial education* refers to educational practices in the context of any colonial encounter. As such, colonial education includes informal and formal schooling practices in French New Orleans in 1727, in Los Angeles in 1825, or in eighteenth-century South India. The term *early national,* as defined here, refers to the period in which the various cultures and social groups within specific geographic regions renegotiated and redefined their identities and social relations as members of a new political entity.

The chapters in this section analyze the influence of the state on national and local policies and the ways that ethnic groups experienced, resisted, and in some cases influenced such policies. To varying degrees, they attempt to consider culture in flexible ways: as fluid, contested, and rooted in traditions and practices as well as beliefs and values. This definition of culture stands in contrast to older, more homogeneous views in which various social groups have often been cate-gorized on the basis of their distinct—and presumably static—"cultures."

Perspectives on the Southwestern Latino School Experience, 1800–1880

As Victoria-Maria MacDonald and Mark Nilles demonstrate, the various cultural groups in the American Southwest experienced several periods that might be conceived respectively as "colonial" or "early national" with regard to schooling: the late eighteenth and early nineteenth centuries under Spanish rule, the early nineteenth century as part of the Mexican Republic, and the mid-nineteenth-century transition to statehood in the United States.

Drawing on legislative reports from the Spanish, Mexican, and U.S. govern-ments; annual reports of schools; and letters and documents from missionaries, teachers, and other involved parties, MacDonald and Nilles provide a richly detailed picture of Latino educational experience in the early nineteenth century. Their work explores the contrast between stated official roles and governmental ideals of public schooling and the actual forms of schooling as they evolved in practice. Additionally, their study challenges traditional understandings of what constitutes "colonial" and "early national" schooling.

Cultural Categories, Hegemony, and the Schooling of the Lumbee Indians in Nineteenth-Century North Carolina

Heather Kimberly Dial explores issues of hegemony and culture from the historical perspective of the North Carolina Lumbee Indians. Today, the Lumbee is a tribe of nonreserved, nonfederally recognized Indians and the largest nonfederally recognized tribe east of the Mississippi, with a legacy of mysterious and unclear origins. As such, they experience a continuous need to convince others of their identity, particularly in their home state of North Carolina. As a member of the Lumbee community, Dial brings to her study of Lumbee history a keen understanding of its contemporary consequences.

In her review of the literature, Dial surveys a wide range of secondary sources to investigate the historical experience of the Lumbee as a specific cultural group living in a society that acknowledged and recognized only two categories: "black" and "white." Drawing on new theoretical frameworks from the field of historical anthropology, she shows how dominant groups have used such restrictive and static categorizations to relegate the Lumbee to a nonentity status, and she argues that the Lumbees' struggle for recognition as Native Americans was linked to their quest for their own schools during the time of segregation.

Dial argues that the Lumbee culture has been shaped by early European contact and the Lumbees' subsequent adaptations for survival and success. For the Lumbee, culture is both a unifying aspect that binds them as a tribal people and a limiting categorization that has been used by the state to deny them recognition as a tribal people.

Conclusion

Reflections on the Historicality of Education Systems and the State

What factors contribute to transformations in schooling? While some theoretical concepts in the secondary literature have sufficient power to explain the range of case studies in this book, others do not. One of the benefits of the case-study approach is that it allows the researcher to test a broad theory against the historical development of social processes in a specific context. The chapter begins by analyzing the diverse case studies in this book in light of the following concepts from the secondary literature: (1) the role of origins in influencing the evolution of centralized or decentralized systems of schooling and (2) restriction, substitution, and cultural incorporation as factors in educational transformations. Several of the authors in this book identify additional factors in educational transformation, including: (1) the influence of international policy networks, and (2) the co-option of market-based structures and processes by the state.

Taken as a whole, the chapters in this book suggest that social inequalities can persist in both highly centralized and decentralized systems. In highly centralized systems, groups with the greatest political power can prevail over others in establishing educational structures and processes that best meet their own class interests. In highly decentralized systems, subordinate groups may succeed in establishing alternative forms of schooling through acts of substitution, but such acts can have the unintended consequence of ultimately reinforcing the hegemony of more powerful groups. In all systems, power is always contested and recreated, but the outcomes of such interactions are far from predictable.

This chapter concludes by arguing that educational systems are more changeable than has been portrayed in the past. Over time, highly decentralized systems can become more centralized in the face of financial restraints or in response to internal or external political pressures; highly centralized systems can become decentralized to accommodate the culture of subordinate constituents or the class interests of dominant groups. This lack of continuity has been obscured by a traditional focus on large, state-funded education systems in Western states, a focus that has produced an illusion of long, enduring historicality.[12]

Notes

1. For an overview of this argument in the context of schools in the United States, see Milton Friedman, "The Role of Government in Education," in *Economics and the Public Interest*, ed. R. A. Solo (New Brunswick, NJ: Rutgers University Press, 1955), 123–144; John E. Chubb and Terry M. Moe, *Politics, Markets, & America's Schools* (Washington, D.C.: Brookings Institution Press, 1990); Carnegie Foundation for the Advancement of Teaching, *School Choice* (Princeton, NJ: Carnegie Foundation, 1992); J. Merrifield, *The School Choice Wars* (Lanham, MD: Scarecrow Press, 2001); H. J. Walberg and J. L. Bast, *Education and Capitalism: How Overcoming Our Fears of Markets and Economics can Improve America's Schools* (Stanford, CA: Hoover Institution Press, 2003).

2. For an overview of these debates in several Western countries, see Peter W. Cookson, ed., *The Choice Controversy* (Newbury Park, CA: Corwin Press, 1992); Michael Engel, *The Struggle for Control of Public Education: Market Ideology vs. Democratic Values* (Philadelphia: Temple University Press, 2000). For a discussion of the impact of choice policies on urban Catholic schools in Great Britain, see Gerald Grace, *Catholic Schools: Missions, Markets and Morality* (New York: Routledge, 2002). For an analysis of the impact of increased competition among schools in New Zealand, see E. B. Fiske and H. F. Ladd, *When Schools Compete: A Cautionary Tale* (Washington, D.C.: Brookings Institution Press, 2000).

3. Adam Smith, *An Inquiry into the Nature and Causes of the Wealth of Nations*, ed. Edwin Cannan (New York: Modern Library, 1994).

4. *Hegemony* is defined as the ability of dominant groups to shape political agendas and social policy. See Antonio Gramsci, *Selections from the Prison Notebooks* (London: Lawrence and Wishart, 1971).

5. Raymond Williams, *Marxism and Literature* (London: Oxford University Press, 1977), 112.
6. Margaret Scotford Archer, *Social Origins of Educational Systems* (London: Sage, 1979). The quote is on page 2.
7. Andy Green, *Education and State Formation: The Rise of Education Systems in England, France, and the USA* (New York: St. Martin's, 1990). The quotes are on pages 77 and 311, respectively.
8. Margaret Scotford Archer, "The Theoretical and the Comparative Analysis of Social Structure," in *Contemporary Europe: Social Structures and Cultural Patterns,* ed. Salvador Giner and Margaret Scotford Archer (London: Routledge and Kegan Paul, 1978), 1–27. The quote is on page 24.
9. Carl F. Kaestle and Maris A. Vinovskis, *Education and Social Change in Nineteenth-Century Massachusetts* (New York: Cambridge University Press, 1980).
10. Michael Apple, ed., *The State and the Politics of Knowledge* (New York: Routledge, 2003), 221.
11. Ting-Hong Wong, "Education and State Formation Reconsidered: Chinese School Identity in Postwar Singapore," in this volume. The quote is on page 41.
12. The term *historicality* is derived from Andrew Abbot, "The Historicality of Individuals," *Social Science History* 29 (Spring 2005): 1–13.

PART I

Education and State Formation

1

Middle-Class Formations and the Emergence of National Schooling: A Historiographical Review of the Australian Debate

Geoffrey Sherington and Craig Campbell

The shopkeepers include some who appear well to do, and others whose stock-in-trade falls below five pounds in value. The clergymen, lawyers and doctors whose child attend National Schools are not of necessity wealthy men, and of the opulent classes not more than four in a hundred are to be found . . . When a system of checks has been devised to secure thoroughness in all the teaching, thoroughness in the discipline, and thoroughness in the testing of results, it cannot easily be perceived where there is room for pretence. It occurs to me, however, that the efforts of the Board to provide a comprehensive education for children of all the various classes attending National Schools may have provoked the remark that they were attempting more than was necessary, and thereby too much "show". Or it may be that undiscerning visitors to the school, seeing the children have concluded that their parents were rich. So far is this from being the case, that one of the cleanest, neatest and most pleasing is the child of a letter carrier in Sydney.

William Wilkins—Secretary to the National
Board of Education, New South Wales, 1865

In *Education and State Formation*, Andy Green provided an account of the rise of national education systems in England, France, and the United States. Rejecting earlier views based on either a "Whig" view of progress or other more functional or economic explanations, Green has argued that the key issue in the timing and development of education systems is the nature of the state and state formation. Centralized states such as post–1789 France created centralized bureaucracies; decentralized states such as the United States created more decentralized public systems, often based on local communities. Allied to the forms and content of education was the nature of class relations in different national contexts. Green sees the case of England as representing the relative weakness of state or public forms of education. The English retained a "Liberal Tradition" that delayed and then limited state intervention.[1] As a result, England retained a gentrified and antiquated system of secondary education dominated by the English public schools, while more genuine middle-class schools emerged in Europe and the United States.

From such a comparative theoretical framework it is useful to reflect upon the development of "public" educational systems in those colonial societies of settlement that became part of the British Empire. Specifically, we need to understand how concepts of "national" and then "public" education developed in a colonial settler context and how they related to the changing nature of communities and class formations.

The early historiography of Australian education was written predominantly in terms of the changing relationships between the state and the various Christian churches. There was a particular concentration on the administration and financing of educational endeavors. The process whereby the colonial state first supported the efforts of religious denominations only to withdraw financial aid in order to create a public education system administered by a central bureaucracy was seen as a natural evolution. The "free, compulsory and secular" acts, which established public education under State Departments of Public Instruction in all the Australian colonies in the two decades from the 1870s, were analyzed as both a necessary response to the problem of establishing universal schooling in a vast continent with a small population as well as a prescription for the proper role for the state guided by the principles of nineteenth-century liberalism in creating opportunities for all. The expertise of a central state bureaucracy running a system of schools staffed with trained teachers would eventually help create a "ladder of opportunity" for all children, whatever their social background.[2]

The only major opposition to this interpretation came from Catholic historians of education. Instead of seeing the creation of public education in Australia as a natural process, they portrayed the withdrawal of aid for church-run schools as a way of denying social justice to the Roman Catholic community, who made up almost one-third of the nonindigenous population. The emergence of the liberal state was associated with the rise of secularism as part of a general movement in Western society with particular implications for colonial Australians. Rather than public education in Australia being seen as an expression of agreement among the colonial population, the "secular" acts were seen at best as a form of common Protestantism and at worst as a means of proselytizing, to turn Catholic children away from their faith. Instead of participating in the centrally administered and bureaucratic public education system with a lay teaching force, the adherents of the Catholic Church increasingly withdrew to create their own schools based on local parishes and staffed by religious orders.[3]

Such were the two "heroic" and overlapping "myths" that had emerged in the historiography by the end of the 1950s. While these "stories" of Australian education shared some of the issues associated with the emergence of school systems in Britain, Europe, and North America, they also had features peculiar to the Australian colonial past. Whereas in Britain and much of Europe the state continued to support the educational efforts of the churches, in Australia, from the late nineteenth century the difference between public and Catholic schooling was one of the major social and cultural divides. And in contrast to the United States, and even to Canada, where state support for church schools also continued, after a period of some local community involvement from the 1850s to 1870s, the neighborhood public school in Australia came to be provided and controlled by the central state administration, with little regard to the claims of local parent and citizen groups.

By the 1970s, these older interpretations of the history of education were being supplanted. A new generation of historians, influenced in part by the then revisionist and New Left movement in the United States and Britain, challenged the view that the nineteenth-century Australian liberal state had acted in the interests of all social classes. Rather than the creation and development of a public education system that served all, it had divided the society, sustaining differences based on class, gender, and race. Centralization prevailed over local communities in the interests of allowing the development of a capitalist state. The working class, rather than being seen as welcoming the actions of the state, appeared to resist the intrusion of the educational bureaucracy and oppose measures that compelled their children to attend school.[4] There was also the question of the persistence of the racial divide between indigenous and non-indigenous Australians. Some historians pointed out that public education had only been for whites; until the mid-twentieth century, the aboriginal and indigenous populations of Australia were often either excluded from public schools or educated in inferior institutions.[5]

This new view of public education tended to exclude issues associated with the question of state-church relations. But matters concerning religion and culture remained difficult to ignore. Beginning in the 1960s a new generation of Catholic historians drew attention to the close relationship between religion and ethnicity. The nineteenth-century Catholic episcopacy in Australia was influenced not only by the views of the papacy toward the nation-state but also by the changing social and political situation in Ireland, where an austere and authoritarian form of Catholicism had begun to develop by the mid-nineteenth century. The Irish-born bishops in Australia increasingly took a hostile view of public education, seeing it as a form of English imperial Protestantism. With increasing influence over the laity, the Catholic bishops appealed to traditions of faith and culture among a Catholic population drawn overwhelmingly from Ireland.[6] At the same time, other historians have drawn attention to the climate of anti-Catholicism that marked much of the political debate and discussion in the Australian colonies in the 1860s and 1870s, when loyalties divided along sectarian lines.[7]

By incorporating gender into the analysis later versions of revisionist history also gave a new place to religion in an interpretation of state formation and patriarchy,[8] as did forms of post-revisionism in the history of education. On the basis of slight empirical evidence one account challenged the inherent social-control thesis based on class relations that was contained in much of the revisionist agenda. Instead, a new alliance between church and state was presented in which the spiritual guidance of religious pastors molded future citizens in the interests of a new form of social governance. In this account, the Protestant churches at least worked with a state "pastoral" bureaucracy.[9]

The new interest in questions of cultural formation and religion has been matched by a revived interest in the middle class in Australia. The major text on class structure in Australian history has questioned the very existence of a "middle class" when the concept is examined in terms of occupational and economic change.[10] A more recent account suggests that that the middle class is best understood as "a projected moral community whose members are identified by their possession of particular moral qualities, political values and social skills."[11]

It is also important to recognize the relationship between the formation of the Australian colonial middle class and the political ideology of liberalism. An educational agenda was crucial to the British middle-class immigrants of the mid-nineteenth century who helped to carry forward ideals of a free press and universal male suffrage as well as being the leaders of representative government in the colonies. Such men believed in the autonomous, self-sufficient individual acting in a rational and moral way.[12]

More generally, the study of the relationship between the individual and the state has been an ongoing feature of the historiography of Australian liberalism.[13] One recent account suggests that a form of "cultural liberalism" focusing on the autonomy of the individual and his or her right to liberty had emerged in Australia by the late nineteenth century. This cultural liberalism was also associated with a faith in the power of reason and a belief in human evolution and social progress, often through the agency of the state.[14]

This chapter takes the discussion of liberalism back to the early to mid-nineteenth century to understand a generation of men who were still committed to a religious interpretation of the world. It seeks to reinterpret the creation of state-supported education in the mid- to late nineteenth century by examining the views and role of four male middle-class immigrants who arrived during the 1830s and 1840s in New South Wales, the first Australian colony to be set up and where the early colonial forms of intellectual liberalism were established.[15] Over four decades following the 1830s, each of these nineteenth-century liberals played a major role in constructing and redefining the role the state should play in education in a British settler environment.

Visions of National Education

In the three decades following the end of the Napoleonic Wars, the colony of New South Wales was transformed from a penal establishment into a British settler society. Beginning in the second decade of the nineteenth century, government regulations created a new class of emancipated convicts who were granted, or soon acquired, land. Many of the original indigenous inhabitants were displaced as the new settlers pressed inland from the coast. During the 1830s, these "emancipists" had been joined by new immigrant settlers from Britain and Ireland. Chains of migration formed across the seas as the settlers moved into urban and rural areas, bringing with them their cultural and religious traditions. New South Wales soon became a young country with new family formations and high birth rates. By mid-century, couples marrying in the Australian colonies could expect to have five or six children.[16]

The new settlers brought new ideas often formed in the period of discussion of political reform in Britain prior to and in the wake of the 1832 Reform Act. In New South Wales British liberal radicalism was transformed into an engagement with such issues as political authority, land policy, convict transportation (finally ended in the 1840s), religion, and education. All such issues raised questions about the proper role of the colonial state.[17]

The historiography has long recognized that the administrative and military state had been crucial in the early years of the penal colony of New South Wales. Its role included provision for the education of the children of the convicts, undertaken principally by the few clergy of the established Church of England.[18] From the early nineteenth century there was also a variety of private venture schools that catered principally to the small commercial and landed elite in the colony. By the 1820s the educational landscape was very diverse, with a variety of church-supported and private schools. Middle-class academies and grammar schools had even been established in the capital, Sydney.[19]

Much of the early historiography focused not on middle-class education but on the early efforts to provide "schools for the people." Certainly, the transformation of New South Wales into a British-settler society with many different religious faiths challenged the primacy of the established church. During the 1820s, the efforts of the Church of England to form a church and schools corporation supported by land grants foundered on the opposition of both Roman Catholics and other Protestant denominations.[20] This failure created the context for the politics of state support for education that would be played out over the following five decades.

An early solution that was proposed to the problem of providing schools in a settlement of different religious faiths came from Ireland. Richard Bourke, the governor of New South Wales from 1831 to 1837, was an Irish landowner of Whig sympathies. A communicant of the Church of Ireland, he favored Catholic emancipation and was fully aware of the controversies over the establishment of schools in Ireland. He was also a strong supporter of the Irish National System of schools, introduced with state funds in 1831, whereby children of Catholic and Protestant faiths would attend a common school but with provision for access by clergy and priests. Bourke proposed such a scheme for New South Wales soon after his appointment as governor. However, firm opposition from the Anglican bishop of Sydney blocked the proposal.[21] More generally, it has been argued that Bourke's proposals failed because they were essentially a form of liberal paternalism from above that lacked popular support while most of the gentry class and senior officials in the colony were also opposed to them.[22]

The ideal of the Irish National System would continue to influence many middle-class immigrants who arrived in New South Wales following Bourke's departure. Its cause was advanced by Robert Lowe, the future English politician who was later responsible for the introduction of the notorious "payment by results" system for English elementary schools when he was Vice President of the Committee of the Council of Education in the 1860s. A graduate of Oxford University, Lowe came to Sydney as a young immigrant. During his short stay in New South Wales from 1842 to 1850 he became a major force on the Legislative Council of New South Wales, the body created prior to the establishment of a full representative government in 1856. Lowe opposed the effort of the British government to renew convict transportation to the colony. He was a major advocate for constitutional change although he opposed the introduction of universal male suffrage. He also played a significant part in forming government policies toward education. He chaired a select committee of the Council in 1844 that surveyed the provision of schools in New South Wales. The recommendations of the committee would eventually lead to the introduction in 1848 of a Board of National Education to support schools based on the principles of national education and offering a Christian-based but nondenominational curriculum. At the same time, a Board of Denominational Education was established to administer education grants to specific religious denominations.[23]

Support for "national education" was part of what has been described as the emergence of a faith in "moral enlightenment" in the Australian colonies. Growing from the eighteenth-century Enlightenment, this new faith merged elements of early nineteenth-century liberalism and romanticism. Through education in particular, all colonists could become "good, wise, prosperous and responsible."[24] It was a faith that motivated many of the promoters of "national schools."

William Augustine Duncan

Born in Scotland in 1811, William Augustine Duncan migrated to New South Wales in his mid-20s. The son of a Scottish farmer whose family faced financial difficulties following his death, Duncan was a brilliant school student who converted to Roman Catholicism. He took preliminary training to enter the Benedictine Order but soon quarreled with his teachers. In the early 1830s, he became a bookseller and publisher in Aberdeen. With a growing interest in politics, he was a strong advocate of the 1832 Reform Act. When his business failed, he took up teaching and journalism. Learning of and approving Governor Bourke's proposals for national education, he came to New South Wales in 1837 to take up employment as a teacher in one of the first Catholic schools.[25]

After a short stint as a teacher, Duncan became the founding editor of the Roman Catholic *Australasian Chronicle*. As editor and publicist, Duncan championed the rights not only of his church but also of small farmers and workers, and opposed the large landowners and their claims to be a colonial aristocracy.

Against such claims he argued for the growth of representative government.[26] As an erudite scholar and a Scottish convert to Catholicism, Duncan found that he had little in common with even the wealthier members of the colonial Irish Catholic community. As he later wrote in his autobiography, many of the leading adherents of the church were "of the emancipated class and though supposed then to be men of great wealth, were extremely illiterate and to the last degree unprincipled."[27]

Duncan was initially a supporter of J. B. Polding, the English Benedectine bishop of Sydney from 1834 to1877. This support brought him into conflict with Sydney's leading Irish Catholics, who engineered his removal as editor of the *Australasian Chronicle.* Duncan then established his own *Duncan's Weekly Register of Politics, Facts and General Literature,* appealing to the small but growing circle of liberal intellectuals and literary figures in Sydney in the 1840s. He continued to oppose the dominance of narrow class interests, extending his criticism to the new "squattocracy" comprising those who had acquired their large holdings by simply "squatting on" or taking over large parcels of crown land, displacing the local Aboriginal populations. Emphasizing the compatibility between liberalism and Catholicism, and thereby reflecting many of the elements that marked the liberal Catholic movement in Europe in the two decades before the 1848 revolutions, Duncan deplored and opposed those who promoted the alienation of Catholics from the rest of the community. In particular, he took issue with the Roman Catholic Church's opposition to national education.[28]

In 1846, following the closure of his *Register* on financial grounds, Duncan moved to Moreton Bay near the settlement of Brisbane, which was then still part of New South Wales. He now became customs officer for the colonial government. Pursuing his literary interests, he became the founding president of the Brisbane School of Arts, continuing to support and argue for the establishment of schools based on the principles of national education. In 1850, he published a *Lecture on National Education,* the first pamphlet ever printed in Brisbane.[29] The ideas reflected in this pamphlet provide a specific perspective on a representative of liberal Catholicism in the mid-nineteenth century when the Australian colonies were on the verge of a major population expansion following the discovery of gold.

Duncan began his *Lecture* with the assertion that "the subject of Public Education is one, the importance of which has been felt and admitted by the wise and good of all ages and nations."[30] As with other nineteenth-century liberal proponents of universal education, he claims that it is an "undisputable fact" that in those nations where education is "generally diffused" the population is "most industrious" as well as "most orderly in their manners." With his knowledge of history he also asserted that in the ancient past and particularly among the Jews and early Christians, education was not necessarily placed with the "priesthood." The dominant role that the clergy and religious orders came to play in education during the Renaissance was due to the fact that they were the only groups in Europe who possessed the necessary literary education. The predominance of the churches continued, so that lay teachers came under ecclesiastical control even with the assent of individual nation-states. And making an obvious reference to the previous plight of Catholics in England prior to Catholic civil emancipation, he suggested that education became "a monopoly in the hands of a party" while part of the population was debarred from education as a public right, in some instances actually being prohibited from establishing their own forms of education. The result, according to Duncan, was that "public education languished, and had in many places become—and nowhere more than in England and her colonies—an object of contempt."[31]

In the Australian colonies, Duncan argued, there was a further reason for the "miserable state of education." Instead of having an established church, there were four churches (Church of England, Roman Catholic, Presbyterian, and Methodist), all supported by the state. There were thus "*four* kinds of public schools, in which different doctrines are taught at the public expense."[32] The answer lay, he suggested, in adopting a "national education" similar in form to that which already existed in Europe and particularly in Germany, Austria, Holland, and Belgium. In response to those who asserted that national education excluded religion, Duncan affirmed that "no education can be perfect which is not based upon Christianity." And reflecting his own liberal Catholic stance, he replied that the answer lay in the incorporation of the holy scriptures into the curriculum so that "the history and morality of the Bible are completely interwoven with the National system."[33] Moreover, the newly appointed commissioners of education on the colonial National Board allowed parents and pastors to add such doctrinal instruction as they deemed necessary. On the other hand, the composition of the board, involving men of all denominations, would soon be alert to "proselytising" by any teacher in the national system. Already the new board had published two volumes of readers drawn from the Old and New Testaments (these were in effect the imported school texts of the National School Commissioners in Ireland). Any claim that the national system had a tendency to religious "infidelity" was therefore a false parallel with other systems such as the public schools of France.[34]

Conscious of the ongoing debate in Ireland over national schools, Duncan nominated supporters such as Thomas Arnold, the former headmaster of Rugby, who had helped to compile the National Board texts. He also referred to the position of his own church, claiming that he had no fear of the Vatican opposing the system. Even though some Catholic bishops and clergy were strong critics, he could cite a number of leading members of the church as supporters, such as Murray, the archbishop of Dublin and a member of the National Commissioners in Ireland. And in one of his clearest calls for religious harmony, Duncan made a pointed case for national education over the different denominational schools:

> It will surely be admitted that young men who have been accustomed to read these admirable lessons in class together—who, notwithstanding some differences of faith, are yet united in youthful friendships— . . . such persons are in a better disposition of mind for investigating truth than those who, educated in different schools have been accustomed,—as some other children I wot [*sic*] are accustomed—to argue with a heat that may consume rather than enlighten, and whose chief arguments, are the abominable nicknames of Papist or heretic, Puseyite or Puritan.[35]

This was a clear and bold statement for the importance of schooling as a form of moral enlightenment and tolerance. It was a way toward forms of common citizenship in an immigrant society. While framed from a context of "common Christianity" among the various denominations, it marked a clear view of the importance of reaching a consensus through public education in a society that had become religiously, if not culturally, diverse. But Duncan was also aware of the interests of his own social class. He concluded his pamphlet with reference to the gangs of youth known as the "Cabbage Tree mob" (a reflection of the hats they wore) who were then terrorizing the citizens of Sydney: "If you look to your streets (in the new settlement of Brisbane) you will see the numbers of fine children growing up as wild and unfettered as the Aborigines themselves . . . It is admitted that Ignorance is the Mother of Vice."[36]

This remained the major statement of Duncan on the issue of education. He returned to Sydney in 1859 and became a member of the National Board of Education and its successor, the Council of Education. Increasingly, his views on national education placed him in conflict with the leading members of the Catholic Church, particularly the Irish-born bishops with whom he shared so few interests. In some respects, he may be seen as a "cultural fragment" of the Scottish Enlightenment tradition in Australia.

George William Rusden

While Duncan became a propagandist for national education, George William Rusden was one of its early agents. His religious background contrasted markedly with Duncan's. Rusden's father, George Keylock Rusden, was a Church of England clergyman and schoolmaster who migrated to New South Wales in 1833 with his wife and ten of his children, following in the path of his eldest son, who had arrived there three years earlier. George Rusden became a priest at Maitland, north of Sydney; he was a strong supporter of the Anglican bishop William Broughton and a firm opponent of efforts to introduce national education.[37]

On the voyage to Australia, the young George William Rusden met Charles Nicholson, a Scottish medical doctor who was also migrating to New South Wales. Nicholson soon inherited considerable property from his uncle and became one of the wealthiest men in the colony. A liberal conservative, Nicholson became a member of the Legislative Council of New South Wales in the 1840s. He was greatly interested in education and was one of the promoters of the establishment of the University of Sydney in 1850. He became a patron of George William Rusden, first making him the manager of a number of his rural properties. In 1849, after Rusden had spent a period of time in China, Nicholson secured for him the position of agent for national schools.[38]

The task allotted to Rusden was to travel rural New South Wales, then including the districts of Port Phillip to the south and Moreton Bay to the north. His time spent managing the properties of Nicholson had acquainted him with much of the country he had to traverse. He had even developed sympathy for the Aboriginal population, learning bits of the language from the many tribal groups.[39]

In a period of 18 months, Rusden travelled almost 10,000 miles, visiting many small hamlets as well as some of the emerging rural towns. On occasions he met opposition from the local clergy, who had established their own denominational schools. He reported back to the national commissioners in Sydney on areas where he thought national schools could be established and could prosper, concluding that the success of such schools would depend upon the support from prominent citizens as school patrons, and in rural areas this essentially meant the local squatters.[40]

In 1851, Rusden became a clerk in the colonial secretary's office of the new Victorian government. The governor of the new colony now transferred the national schools in the former Port Phillip district to the Denominational Board. The commissioners on the Denominational Board in Victoria recommended that any schools founded on the national system should be only established in rural districts and then only if the area did not contain a large majority belonging to one denomination. It was also proposed that a General Board of Education rather than a specific National Board of Education should administer these schools.[41]

This proposed action stimulated Rusden to compile his own pamphlet, on *National Education,* writing the work over a period of 12 months. The 365-page book was published in early 1853. Rusden's biographer has suggested that the publication had been hastily researched and composed, but it remains a compendium of a colonial account of national education not only in Australia but also in Britain, Europe, and North America.[42] In putting forward his case for a state-supported system of national schools, Rusden drew upon a number of historical and philosophical texts. Significantly, the book was still framed from the continuing commitment of Rusden to the Church of England. The initial chapter entitled "Patriarchial Education," suggested that formal instruction and learning formed part of Jewish culture and is justified in biblical texts.[43] After outlining the educational traditions under the ancient Persians, Greeks, and Romans and in medieval Europe, most of the book is concerned with a review of national education systems in early nineteenth-century North America, Britain and Europe. Rusden was particularly aware of the developments in Germany and North America. The work of Horace Mann in Massachusetts brought strong praise from Rusden. He also outlined the recent developments in state-supported education in both Ireland and England, relying for the English situation on the words of H. E. C. Childers, the inspector of denominational schools in Victoria.[44]

It is only in the second half of the book that Rusden provides a detailed justification for the introduction of national education in Australia. Essentially, he argued that the national system was necessary because the experience of the 1830s and 1840s had shown that the denominational system had failed to extend schools to all children. Drawing upon his own experience, he suggested that the principles of national education are such that it must provide for all:

> Principles to be good, must be everywhere applicable. The advocates of National Education contend for Christianity as zealously as any one, and can at the same time conscientiously do what Denominationalists cannot viz., they can recommend

the formation of a sound Christian National School in any part of the colony, even though not one family of the persuasion of any particular advocate of the National System may reside in such parish.[45]

Much of Rusden's defence of national schools was designed to disarm the critics within the Church of England. Like Duncan, he relied upon the support for national education in Ireland with particular justification drawn from the views of such supporters as Thomas Arnold. He also denied the claim that denominational schools had an advantage over national schools in terms of local governance. Rather, he saw the local patrons and local boards in the national system as enshrining the English principle of local government, compared to the "mongrel system" of local representation in the denominational system, where the clergy guided the local board and the bishop controlled the board with power to remove the teacher.[46] Moreover, he asserted that his own experience in touring the colony indicated that any success he had achieved was due to the cordial support he received. "The public joined in agency for the National System:—by their efforts only was it (so far as my personal experience goes,) sanctioned at every meeting convened for the purpose of discussing the matter."[47]

To Rusden, national schools were thus justified on grounds of both efficiency and liberal democracy. Drawing on the experience of the New England States and the ideal of the "common school," he argued that "where Governments support comprehensive or National Education, the mass of the community attains a higher intellectual position than has ever fallen, or than is likely to fall to the lot of countries where sectarian Education is supported at the cost of the public treasury . . . what country can be found, which, giving Sectarian education, can vie with the New England States of America in the results of its Educational labors?"[48]

The knowledge of and reference to the United States was particularly significant in terms of Australia as a British-settler society. While much of the nature of the colonial national schools relied on the Irish experiment, Rusden revealed here a fascination with the common school ideal, emphasizing (with an apparent limited understanding of the American experience) that all "National Schools" were not "secular" but rather still taught Christianity, constantly drawing upon the gospels.[49]

In his final and concluding chapter, Rusden clearly laid out his vision for a unified and state-supported system of schools that would respect multiple faiths and religious convictions:

An advocate for such a system is impelled by no hostility to sects, and by no rapacity for self;—he contends for the broad principle that there shall be schools aided by public money; that all children shall be able to claim access to such schools; and that in them there shall be such united or combined Education, as may induce good fellowship amongst the subjects of the State, while at the same time there shall be no teaching,—no authorized regulation—doing violence to the consciences or religious convictions of children or of parents;—thus, therefore, the common funds of the nation are held sacred for public, not for sectional purposes; but free facilities remain by which the sectional predilections of different communions may be gratified, as indeed they ought to be, if at all, at the cost of the sects themselves.[50]

The implication of this vision would be to provide state funds only to common national schools, leaving the various denominations to maintain their own schools should they so wish. To Rusden, this was an indication of the liberal state in action, for "if the state is bound to confer education, it is plain that the obligation is a general one," as society was framed not to cater to individual interests but to regulate affairs involving general principles affecting the whole community.[51] To implement this principle of state-supported common schools, Rusden suggested that there be "school-rates" as had operated in North America, providing for a minimum amount of school provision in each district, leaving voluntary efforts to do as "much more as benevolence may prompt." There was also compulsion in his vision. He would even require a "double school rate" from parents declining to send their children to school while at the same time each school district would be required to offer "gratuitous education to those few children whose parents or guardians might really be unable to discharge their required duties."[52]

Rusden's vision of compulsory universal national education predated the free, compulsory, and secular acts by two decades. He had moved the debate on from merely an effort to have the national schools accepted toward the view that the state should only provide aid to a system that could embrace all children involved in a common curriculum. It is not known how many copies of *National Education* were sold, but it evoked considerable interest and received a number of favorable reviews. Many of the arguments for national education now rested on the views expressed in the book.[53] Rusden himself gave evidence to the 1852 Select Committee on Education for the new colony of Victoria, claiming that the best "public schools" were the national schools then in Sydney, and urging the creation of a central authority capable of establishing and operating public schools for Victoria.[54]

Despite this attention to his views, Rusden himself grew increasingly disillusioned with the educational and political developments that now occurred. While he was a member of the Board of National Education in Victoria from 1853 to 1862, he faced continuing opposition from those who supported the existing denominational schools as well as those who advocated a "secular" curriculum rather than one based on common Christianity. In 1862, a Common Schools Board was created to administer funds to all schools. Rusden, who was not appointed to the new board, increasingly found himself out of step with those who saw the only way forward being the introduction of a secular form of state schooling rather than the vision of common agreement among the various Christian religious denominations.

Creating a Public School System

By the 1860s, national schools were a well-established part of the educational landscape of New South Wales. Over the next two decades, the colony moved from a "dual" system of state-supported national and denominational schools under two separate boards toward one in which public schools funded by the state and administered by a central bureaucracy emerged. The process whereby this occurred is related to two pieces of colonial legislation—the Public Schools Act of 1866 and the Public Instruction Act of 1880. These legislations have also long been associated with the role of William Wilkins as educational administrator and Henry Parkes as colonial politician. The role of these two men in the administration and politics of education was part of a cultural consensus that emerged among colonial middle-class Protestants in the mid- to late nineteenth century.

William Wilkins

William Wilkins was a creation of the state-supported educational systems that emerged in Britain and Australia during the nineteenth century. Born in 1827 in London in the Lambeth Workhouse Infirmary, he was first schooled in the Norwood School of Industry. In 1840 he became one of the early pupil teachers at the Battersea Training School under James Kay Shuttleworth. He then became assistant schoolmaster at Parkhurst Prison on the Isle of Wight and later worked as a teacher in Manchester and St. Thomas National School, Charterhouse. In 1850 he was offered the position of headmaster of the Fort Street Model School, Sydney, established by the new National Board of Education to provide training for teachers. By the mid-1860s, as one of the best-known public servants in New South Wales, Wilkins had achieved full middle-class respectability, attending his office in "frock coat, tall hat, gloves, and walking stick."[55]

Wilkins success was due much to the part he played in building the national school system in Australia. He had brought with him his knowledge of the pupil-teacher method and the educational views of Kay Shuttleworth, which were in part influenced by Pestalozzi. There was also the general influence of Kay Shuttleworth as a disciple of Jeremy Bentham and utilitarianism.

In 1854, the Board of National Education appointed Wilkins inspector and superintendent of schools, allowing him to introduce a series of teaching standards. He then headed an inquiry into education in the colony that condemned the inefficient practice of having dual boards of denominational and national education. Instead, Wilkins proposed a unified coordinated system of state schools, supervised by professional district inspectors. He also recommended the extension of the pupil-teacher system and improvement of school buildings. He adopted this agenda in his administration of national education over the following decade, becoming acting secretary to the Board of Education in 1863 and permanent secretary in 1865.[56]

Initially, Wilkins had to defend the existence of the national school system against its critics. Carrying on the vision established by Duncan and Rusden, he maintained that the national school system was not only more efficient than the schools under the denominational board but also both religious and respectable. As such, the intersection between schooling and social class became a more permanent feature of discussion and debate.

In 1857, Wilkins and a number of national schoolteachers published a series of letters in the press to counter a published sectarian (Church of England) attack on national schools. While some of the arguments presented mirrored the earlier views of Duncan and Rusden, what is also notable is the way that social class had now entered the debate. While Rusden had sought the support of the "squattocracy" as patrons of national schools in rural districts for children of the rural laboring class and Duncan had emphasized that national schools were essentially for the poor, Wilkins and his co-correspondents pointed out that in Sydney at least, 2,000 children were now attending national schools: "They are of all denominations, and among them their parents may be found representatives of all classes of society—*ministers of religion,* merchants, professional gentlemen, tradespeople, mechanics and labourers."[57] Thus, it could not be expected that such parents would send their children to "infidel schools," as had been alleged against the national schools.[58]

On the other hand, Wilkins and his teachers rejected the claims that "in these schools, *instead of the poorest,* the middle, and even some of the highest classes are found." Rather, it was the case that there were many of the poor, albeit those who were respectable in appearance and not dirty. Those not in national schools had excluded themselves because they had disregarded the rules of the school requiring punctuality, regularity, and cleanliness.[59]

Such techniques of control and order were part of what has been described as the emerging "liberal classroom." Under the guidance of Wilkins, the teacher replaced the patron as the central moral authority in national schools. In contrast, the denominational schools still remained under the surveillance, if not the supervision, of the local clergy or priest.[60] And, as Wilkins and his colleagues claimed, parents were now looking toward the "best" school, irrespective of religious affiliation: "Once within the walls of the school-room, all pupils are alike to us. We know no distinction of class, sect, or wealth. We hold that the children of every citizen, whether high or low, rich or poor, are entitled to the best education that the state can attend, and none have superior rights over others."[61]

Trained teachers and regular inspection became the hallmarks of the national schools by the early 1860s. The National Board of Education also had an increasing advantage over the Denominational Board. The establishment of parliamentary representative government in 1856 initiated an era of political factions rather than political parties. With governments changing rapidly, instability reigned.[62] What became constant was increasing accountability for the expenditure of public funds. With its growing central controls, the National Board was able to claim efficiency in public expenditure. In contrast, the Denominational Board tried to establish its own inspectorial system but also had to accommodate the various interests of the different denominations to which it disbursed public funds.[63] In 1865, Wilkins delivered two lectures on *National Education* as a way of providing an exposition of the national system of New South Wales. He founded his lectures on three related principles: first, that the state "being in theory the embodiment of the will of the whole people" could only use public funds for "universal benefit"; second, that as such, all institutions supported by the state must be of a "civil character"; and third, that the state cannot interfere in matters of opinion that did not directly affect the obedience of citizens to the law. On these principles, the state could not devote resources to serve only a portion of its citizens. To do this would mean that "schools would cease to be civil institutions."[64]

Having dismissed any justification for the existing state support for denominational schools, Wilkins devoted the remainder of the lectures to the justification of the national school system. He admitted that the problem facing the colonial state was the "Religious Difficulty" or "Ecclesiastical Difficulty." But this difficulty resulted from "fallacies" about the perceived duty of the teacher, the function of the school, the relation of the school to the state, and the nature of religious education as distinct from "dogmatic theology," for all children. As such, much of his emphasis was on the proper function of the teacher and the curriculum in national schools, which was to "teach the elements of secular knowledge" and to inculcate moral and religious truths but not doctrinal theology. As with Duncan and Rusden, Wilkins reminded his audience that this practice of teaching "religious education" and the elements of common Christianity were found not only in the national system in Ireland but also in Europe—specifically, Germany, Belgium, and Holland—and particularly in the United States, where the "principle of combined secular and general religious teaching has been most fully practised."[65]

While emphasizing the significance of a common secular instruction combined with common Christianity, Wilkins once again reasserted that the public school was for all social classes. He rejected the view that only the rich sent their children to national schools. Indeed, as "class legislation is a political wrong," state-supported schools must be open to all. "Properly speaking, the state knows of no class distinctions—all are simple citizens, enjoying equal rights and immunities."[66] Thus, it would be wrong to exclude any citizen from public schools because they had wealth. On the other hand, Wilkins suggested, with a clear reference to working-class Catholics, if the poor were avoiding the national schools, it was probably because of the influence of the clergy. By citing figures to show that the national schools were not biased toward the rich, he claimed that in Sydney there were many small shopkeepers, as well as clergymen, lawyers, and doctors, whose children attended national schools but who were not men of wealth. The "opulent classes" represented only about 4 percent of total enrolments. Overall, the main principle was to hold to the view that a "mixture of all classes in public schools is desirable."[67]

In terms of the future administration of public funds, the colonial government soon acted to overcome what was seen as the inefficiencies of maintaining two separate boards. The Public Schools Act of 1866 created a single administrative body to oversee all "public education." Effectively, this blurred the former distinction between the national and denominational schools, with the Council establishing a system of inspection to ensure that all schools reached appropriate standards. The act also provided for the expansion of what were now known as public schools that could be established in any locality where 25 pupils could attend regularly. In contrast, a denominational school could receive public funds only if it had an enrolment of at least 30 and the local public school not less than 70 children.[68] In practice, many denominational schools now found it difficult to survive, and to meet the new standards of the inspectors of the Council of Education. Even Catholic parents drifted toward the better-resourced public schools.[69]

The Public Schools Act of 1866 and the subsequent regulations of the Council of Education may be regarded as a triumph for the policies of Wilkins. He drafted much of the legislation and saw that the resultant regulations were enforced. But the act also changed the politics of education and sectarianism. The final act in the culmination of public education lay in the hands of colonial politicians.

Henry Parkes

Henry Parkes was perhaps the best known of all Australian colonial politicians. He was born in Warwickshire in 1815, but his family moved to Birmingham after his father was forced off his farm due to debt. By 1837, Parkes had served his apprenticeship as a bone and ivory turner and had married. The early influences on his religious and political views were congregationalism and Thomas Atwood's radical Political Union, which pressed for further political change in the wake of the 1832 Reform Act. Following a business failure, Parkes and his wife came to New South Wales as assisted immigrants. Parkes soon became involved in literary and political affairs in Sydney, with Duncan as one of his close friends and advisers. He urged the universal male franchise and an end to convict transportation, becoming editor and proprietor of the *Empire.* Over the following decades, his fortunes waxed and waned. Bankrupted on a number of occasions, he had to depend for his livelihood on government employment, such as becoming a colonial immigration agent in Britain or a paid minister of the Crown. In the faction politics that followed the creation of parliamentary government, Parkes was at the center of those opposing the ruling establishment of liberal politicians. In 1865, he became colonial secretary in a coalition ministry involving various interests and opinions.[70]

As colonial secretary, Parkes was responsible for the passage of the 1866 Public Schools Bill through parliament. While his own schooling had been limited, he had developed strong views on the importance of education in a colonial democracy. In some respects, his vision was still framed by the earlier views of Rusden and Duncan rather than by Wilkins's understanding of how colonial public education had developed. Parkes's speech in introducing the bill was founded on the principle of the government providing for "popular" education in such manner as had occurred in Holland, Germany, and the United States.[71] Parkes had already taken great interest in the fate of juvenile and delinquent youth, supporting reformatories and industrial schools. He now asserted that it was the "duty of government" to reach the estimated 100,000 "destitute and neglected children" who were without any instruction.[72] Thus, while the new legislation establishing the Council of Education still provided for school fees, Parkes emphasized, in contrast to Wilkins, that if necessary the state would pay for those children who had "disreputable parents": "We provide education for a class of children who require it most, because if any child requires the intercession of the State, if any child has a claim upon its bounty, it is the poor child whose parents neglect to provide for him and refuse to pay the oridinary school fees."[73]

For Parkes, the aim of universal provision still prevailed over differences of culture and religion. Although he had developed a deep suspicion of Roman Catholicism during his years in Birmingham, he accepted the 1866 compromise whereby state funds would continue to be provided for denominational schools. This did not prevent him from exploiting Protestant fears roused by the assassination plot on the life of the Duke of Edinburgh when he was visiting Sydney in 1868. In a famous speech at the regional center of Kiama, south of Sydney, a settlement where Irish Protestants were prominent, Parkes even alleged that the potential assassin had acted on the orders of Irish Fenians and that one conspirator had been murdered lest he reveal the plot.[74]

Despite this, from the mid-1860s to the mid-1870s, Parkes opposed and resisted efforts in the colonial Parliament to remove public funding support for denominational schools. His change of mind over this issue has been attributed in part to his increasing realization that the administration of public funds for education should come only through a state department under a minister responsible to Parliament rather than a council comprising representatives of the various religious denominations as well as supporters of public education.[75]

Reinforcing the views of Parkes was the changing climate of both Protestantism and Catholicism in New South Wales. By the 1870s, the Church of England in New South Wales had come to accept the loss of its former exclusive position. The ecclesiastical leadership now drew on evangelical traditions that found common ground with other Protestant denominations. The church still maintained and defended the role of its own schools but also found other outlets such as Sunday schools as a way of promoting doctrinal faiths.[76] The other major Protestant churches had virtually abandoned their schools in the wake of the 1866 Act and were supporting the public schools. For example, Henry Parkes's son-in-law, who was a Presbyterian minister, worked hard to establish public schools on the south coast of New South Wales. By 1869, Parkes's daughter was even writing to her father for copies of the public school regulations, wanting to know if they could start two half-time schools with one teacher between them.[77]

There was also an example of united militant Protestantism emerging. In 1874, James Greenwood, a Baptist minister, and Zachary Barry, an Anglican clergyman of Irish Protestant background with "Orange" sympathies, formed the Public Schools League, the mainfesto of which became "Education—National, Compulsory and Secular and Free." The actual slogan may have been drawn from the example of the National Education Union that Joseph Chamberlain had formed in 1869 as a prelude to the 1870 Act in England. But the major implication in colonial New South Wales was to give political voice to those Protestants who wanted a new educational settlement despite the formal opposition from the Anglican bishop of Sydney and the Church of England Schools Defence Association.[78]

Paralleling organized Protestantism was the growing strength of the Catholic bishops. From the late 1860s, the Catholic episcopacy throughout Australia, disturbed at the drift of Catholic children to nondenominational public schools, had begun to issue a series of directives to the lay members of the Catholic Church warning them against schools in which the Catholic faith was not taught. In South Australia, the Catholic Church had already begun to abandon even denominational state-supported schools with lay teachers in favor of creating a system of schools under religious orders.[79]

The arrival of a new Catholic archbishop of Sydney in 1878 provided an impetus to the church in New South Wales. In 1879, along with the other bishops of New South Wales, the archbishop issued a famous pastoral letter as a call and warning to the Catholic faithful:

> We, the Archbishop and Bishops of this colony, with all the weight of our authority, condemn the principle of secularist education, and those schools which are founded on that principle. We condemn them, first, because they contravene the first principles of the Christian religion; and secondly, because they are seed-plots of future immorality, infidelity, and lawlessness, being calculated to debase the standard of human excellence, and to corrupt the political, social and individual life of future citizens.[80]

Interpreted as an attack on the public schools of New South Wales, these words provided the opportunity for Parkes, who was now colonial premier, to introduce legislation repealing the 1866 Public Schools Act, effectively removing public funds from all denominational schools and creating a public school system under a State Department of Public Instruction. It has often been suggested that the 1880 Public Instruction Act enshrined Henry Parkes as the father of public education in New South Wales, even though his commitment to its principles seemed less important than his image as a champion of educational change.[81] As he wrote to his daughter following the passage of the legislation: "The good fortune does not fall to the lot of many men to be the acknowledged author of two great measures like the Public Schools Act of 1866 and the Public Instruction Act of 1880—to be in fact the founder of the Primary School system of a country. You must pardon this little piece of private jubilation."[82]

In effect, the principles of the legislation revealed how the visions of national education, first conceived in the 1830s as an answer to the problem of providing state support for "popular education," had now been transformed into a form that was in part inclusive, while effectively excluding a significant section of the colonial population. The Public Instruction Act provided for "non-sectarian teaching" but "secular instruction" meaning "general religious teaching" which, for all intents and purposes, now meant common Protantism. Technically, four hours in each day were given to "secular instruction exclusively," with provision for the clergy to offer one hour a day at least one day a week. Aid to all denominational schools would cease from the end of 1882.[83]

While these clauses on "secular" education in the Public Instruction Act of 1880 have often drawn most attention, it should also be noted that the legislation extended the nature and meaning of public education in New South Wales. In his autobiography, Parkes himself later drew attention to the manner in which he introduced the legislation, indicating that the act would provide not only a new settlement of the question of "secularism" but also an extension of the nature of public education:

> So far as the Bill will make new provisions to supply the wants of education, it will provide for the immediate establishment of grammar schools in three of the principal towns, with provision for the extension of this higher means of education to other districts on proclamation. It will also provide for the establishment of one or more higher schools for girls, to be extended as the circumstances of the population may warrant.[84]

The idea for what became public "high schools" did not come from Parkes. Rather, it originated from his friend and political ally William Charles Windeyer, an early graduate of the University of Sydney and long-standing member of the council of the Sydney Grammar School, established in 1854 with state subsidy as a feeder school to the university. Windeyer had a strong interest in extending education for both males and females, having previously introduced legislation that would have created state grammar schools in rural New South Wales.[85]

The provision for public high schools continued the process begun under Wilkins in the 1850s of making public education attractive to the respectable Protestant middle class. Within a few years, single-sex high schools had been established in Sydney, even though they still struggled to survive in the country areas.[86] The Public Instruction Act also allowed for superior public schools that in many areas would become a form of middle-class schooling offering a curriculum that allowed matriculation to the university.[87] In these ways the Act of 1880 allowed the development of public education to first emulate and then challenge the mainly church corporate secondary schools that had begun to emerge in New South Wales by the late nineteenth century.[88] By the late nineteenth century, public schools were thus both a "popular" and "respectable" form of education. Even the principle of free public education was deferred for another quarter of a century in the interests of at least maintaining the image of citizens making some contribution to their children's future.

The integration of culture and social classes in a "liberal democratic" society formed of immigrants was thus the final justification of the Act of 1880. As Parkes informed Parliament in 1880:

> We think this Bill may be fairly accepted by all—by every class, by every sect. It does not matter whether the child belongs to an Irish, a Scotch, an English, or an Australian family. What is aimed at is that he should be considered as belonging to a family forming part of the population of this free and fair country . . . We think another advantage in this Bill is that it is not a Bill for the poor alone. It is not a Bill conceived in any sense of helping only those who cannot help themselves; but it is a Bill framed and intended to bring into existence a system of education for all the children of all classes; so that the child of the poor and the child of rich may sit side by side in their tender years, when they receive the first rudiments of instruction, and when there is no occasion for any sectarian distinction . . . And I venture to say that they ought so to mix; that they ought to unite in promoting the general interests of their own country in preference to any

other considerations whatever. Let us be of whatever faith we may, born on what-ever soil we may, reared under whatever associations we may, let us still remember that we are above everything else free citizens of a free commonwealth.[89]

Some Conclusions

The early nineteenth-century discussion in Australia of "national education" drew heavily upon the British, European, and North American experience. While the concept of a "national system" often referred specifically to the Irish situation, there was a wider understanding that a new vision was required for the settler societies now being formed in the Australian colonies. In the view of Duncan and Rusden, state action was necessary to support an educational provision for "the people" in ways that would respect the faiths of different communities while allowing the emergence of an Australian version of the common school. The argument in favor of "national education" was made on grounds of efficiency but also in terms of universal enlightenment.

The public schools that grew out of these early visions soon developed particular educational forms. A utilitarian emphasis on teaching expertise and proper forms of administration replaced the earlier reliance on communities to organize their schools. The national schools under Wilkins prevailed over the denominational sector because they could demonstrate that centralization was an efficient way to control public expenditure. At the same time, the conception of public education changed as urban middle-class patronage provided a new respectability for public schools.

As the previous historiography has suggested, administrative centralization was a major feature that marked developments in Australia. This, it has been suggested, distinguished the organization of the Australian public school from that of Britain and North America. Nevertheless, much of the development of the concept of "public education" in Australia seems to have mirrored the American, if not the British, experience. As Bill Reese has suggested, despite the early nineteenth-century Massachusetts model of the "common school," the idea of "public education" and the "public school" only became clear in the late nineteenth century. By 1873, an American public school could be described as one "established by the State through agencies of its providing, conducted according to the rules of its authorization, supported by funds protected or furnished by its legislation, accessible to children of all citizens upon terms of equality, and subject to inspection as the law may institute."[90] This was also a reasonable description of public schools in Australia.

There was a further association between the American and Australian experience. As in Australia, the Protestant public school establishment in the United States was anxious to ensure that particular religious and cultural forms prevailed in a society undergoing change on account of the influx of new immigrants.[91] Indeed, even centralization as the solution to organizing Australian public schools was associated with the emerging Protestant middle-class ascendancy in both state administration and the politics of education. Astute politician that he was, Henry Parkes was certainly conscious of measures to build a nation composed of the white settlers in Australia. He remained committed to the 1880 Act as providing both equality of opportunity and social integration of all communities (with the notable exception of indigenous Australians, whom Parkes and his generation assumed were doomed to extinction). An advocate since the 1850s of the union of all the Australian colonies, by the end of the 1880s he had become the leading figure calling for an Australian federation that was finally established in 1901, five years after his death.[92]

Religion, culture, and social class thus mingled in various ways in the transformation of a vision of national education into a public school system. But this transformation was not founded on cultural or class consensus. As in the United States, the Catholic Church set out to create its own different school systems under the religious orders. If the outcome of the Australian colonial debate ensured the ascendancy of the Protestant middle class, it also demonstrated the weakness of the Catholic middle class. The views of liberal Catholics such as Duncan were marginalized, while the authority of bishops and clergy was strengthened. For almost a century, Catholic parents and their communities had to find ways to build and support their own schools without the promise of state aid. Catholic education in twentieth-century Australia developed its own forms—authoritarian in pedagogy, infused with an all-pervasive "religious atmosphere," promoting specific gender roles, and maintaining a specific Irish influence through the religious teaching orders.[93]

More fundamentally, sectarianism, as much as social class, became the great divide in Australian society. As has recently been argued, the links between Protestantism and middle-class liberalism run deep in Australian society.[94] The development of public schools in the nineteenth century was but one expression of that phenomenon. It was a legacy that public education would carry forward into the twentieth century.

Notes

The authors acknowledge support from the Australian Research Council.

1. A. Green, *Education and State Formation: The Rise of Education Systems in England, France and the USA* (London: Macmillan, 1990).
2. S. H. Smith and G. T. Spaull, *History of Education in New South Wales (1788–1925)* (Sydney: George B. Philip and Son, 1925); G. V. Portus, *Free, Compulsory and Secular: A Critical Estimate of Australian Education* (London: Humphrey Milford, 1937); A. G. Austin, *Australian Education 1788–1900: Church State and Public Education* (Melbourne: Melbourne University Press, 1961).
3. U. Corrigan, *Catholic Education in New South Wales* (Sydney: Angus and Robertson, 1930); R. Fogarty, *Catholic Education in Australia 1806–1950,* 2 vols (Melbourne: Melbourne University Press, 1959).
4. I. Davey, "Growing Up in a Working Class Community: School and Work in Hindmarsh," in *Families in Colonial Australia,* ed. P. Grimshaw et al. (Sydney: George Allen and Unwin, 1985); P. Miller, *Long Division: State Schooling in South Australia* (Adelaide: Wakefield Press, 1986); M. Vick "Class, Gender and Administration: The 1851 Education Act in South Australia," *History of Education Review* 17, no. 1 (1988); M. Vick, "Schooling and the Production of Local

Communities in Mid-Nineteenth Century Australia," *Historical Studies in Education* 6, no. 3 (1994).

5. J. J. Fletcher, *Clean Clad and Courteous: A History of Aboriginal Education in New South Wales* (Sydney: privately published, 1989).

6. P. O'Farrell, *The Catholic Church in Australia: A Short History* (Sydney: Nelson, 1968); P. O'Farrell, *The Catholic Church and Community in Australian History* (Sydney: University of New South Wales Press, 1985); P. O'Farrell, *The Irish in Australia* (Sydney: University of New South Wales Press, 1986); J. Maloney, *The Roman Mould of the Australian Catholic Church* (Melbourne: Melbourne University Press, 1969); G. Haines, *Lay Catholics and the Education Question in Nineteenth Century New South Wales* (Sydney: Catholic Theological Faculty, 1976); E. Campion, *Rockchoppers: Growing Up Catholic in Australia* (Sydney: Penguin, 1982); T. O'Donoghue, *Upholding the Faith: The Process of Education in Catholic Schools in Australia, 1922–1965* (New York: Peter Lang, 2001).

7. M. Pawsey, *The Popish Plot Culture Clashes in Victoria 1860–1863* (Sydney: Studies in the Christian Movement, 1983).

8. P. Miller and I. Davey, "Family Formation and the Patriarchial State," in *Family, School and State in Australian History,* ed. M. Theobald and R. J. W. Selleck (Sydney: Allen and Unwin, 1990).

9. I. Hunter, *Re-thinking the School: Subjectivity Bureaucracy and Criticism* (Sydney: Allen and Unwin, 1994).

10. R. W. Connell and T. H. Irving, *Class Structure in Australian History* (Melbourne: Longman Cheshire, 1992).

11. Judith Brett, Australian Liberals and the Moral Middle Class: From Alfred Deakin to John Howard (Melbourne: Cambridge University Press, 2003), 6–7.

12. S. Macintyre, , *A Colonial Liberalism* (Melbourne: Melbourne University Press, 1991).

13. C. Campbell, "Liberalism in Australian History," in *Social Policy in Australia Some Perspectives 1901–1975,* ed. Jill Roe (Sydney: Cassell, 1976); T. Rowse *Australian Liberalism and National Character* (Melbourne: Kibble Books, 1978); M. Roe, *Quest for Authority in Eastern Australia 1835–1851* (Melbourne: Melbourne University Press, 1965).

14. G. Melluish, *Cultural Liberalism in Australia: A Study in Intellectual and Cultural History* (Melbourne: Cambridge University Press, 1995).

15. J. Docker, *Australian Cultural Elites: Intellectual Traditions in Sydney and Melbourne* (Sydney: Angus and Robertson, 1974); Melluish, *Cultural Liberalism in Australia.*

16. Grimshaw et al., *Families in Colonial Australia.*

17. Roe, *Quest for Authority.*

18. J. Cleverley, *The First Generation School and Society in Early Australia* (Sydney: Sydney University Press, 1971).

19. E. Windschuttle, "Educating the Daughters of the Ruling Class in Colonial New South Wales," in *Melbourne Studies in Education 1980* (Melbourne: Melbourne University Press, 1980); M. M. H. Thompson, *William Woolls* (Sydney: Hale and Iremonger, 1980); C. Mooney, "Securing a Classical Education in and around Sydney: 1830—1850," *History of Education Review* 25, no. 1 (1996); G. Sherington et al., *Learning to Lead. A History of Girls' and Boys' Corporate Secondary Schools in Australia* (Sydney: George Allen and Unwin, 1987); G. Sherington, "Public Commitment and Private Choice in Australian Secondary Education," in *Public or Private Education? Lessons from History,* ed. R. Aldrich (London: Woburn Press, 2004).

20. R. Burns, "Archdeacon Scott and the Church and School Corporation," in *Pioneers of Australian Education,* ed. C. Turney (Sydney: Sydney University Press, 1969); A. Barcan,

Two Centuries of Education in New South Wales (Sydney: University of New South Wales, 1988).

21. K. Grose, "William Grant Broughton and National Education, in New South Wales 1829–1836," in *Melbourne Studies in Education 1965* (Melbourne: Melbourne University Press, 1966).

22. J. Cleverley, "Governor Bourke and the Introduction of the Irish National System," in Turney *Pioneers of Australian Education;* H. King, *Richard Bourke* (Melbourne: Oxford University Press, 1971).

23. R. Knight, *Illiberal Liberal Robert Lowe in New South Wales 1842–1850* (Melbourne: Melbourne University Press, 1966); Barcan, *Two Centuries of Education.*

24. Roe, *Quest for Authority.*

25. Duncan, Willian Augustine, *Autobiography,* Mitchell Library, Sydney, A2877; Michael Roe, "Duncan, William Augustine," in *Australian Dictionary of Biography (ADB),* vol. 1, 335–337.

26. Roe, "Duncan".

27. Duncan, *Autobiography.*

28. Roe, "Duncan".

29. Ibid.

30. W. A. Duncan, *Lecture on National Education* (Brisbane, 1850). Copy in Mitchell Library, Sydney.

31. Ibid., 1–5.

32. Ibid., 5.

33. Ibid., 7–8.

34. Ibid., 12.

35. Ibid., 15.

36. Ibid., 21.

37. A. G. Austin, *George William Rusden and National Education in Australia, 1849–1862* (Melbourne, Melbourne University Press, 1958), 1–16; Ann Blainey and Mary Lazarus, "Rusden, William George," *ADB,* vol. 6, 72–73.

38. David S. Macmillan, "Nicholson, Sir Charles," *ADB,* vol. 2, 283–285; Blainey and Lazarus, "Rusden".

39. Austin, *Rusden,* 7–8.

40. Ibid., 50–77.

41. Ibid., 86.

42. Ibid., 87.

43. G. W. Rusden, *National Education* (Melbourne: The Argus, 1853), 1–12.

44. Ibid., 63–78, 126–145. Educated at Cheam School under Charles Mayo, a disciple of Pestalozzi, and a graduate of Cambridge University, Childers came to Victoria in 1850 and was inspector of denominational schools in 1851. He opposed the idea of separate National and Denominational Boards and wanted to see a General Board of Education with power to give or withhold grants. Childers returned to England and later held office in various governments in the period 1860–1886. H. L. Hall, "Childers Hugh," *ADB,* vol. 3, 390–391. See also E. Sweetman, *The Educational Activities in Victoria of H. C. E. Childers* (Melbourne: Australian Council of Educational Research, 1937).

45. Rusden, *National Education,* 231.

46. Ibid., 306.

47. Ibid., 308.

48. Ibid., 310.

49. Ibid., 63–78, 328–329.

50. Ibid., 335.

51. Ibid., 343.

52. Ibid., 345.

53. Austin, *Rusden,* 90–94.

54. Ibid., 100–102.

55. C. Turney, "William Wilkins: Australia's Kay Shuttleworth," in *Pioneers of Australian Education,* ed. C. Turney (Sydney: Sydney University Press, 1992).

56. C. Turney, "Wilkins, William", *ADB,* vol. 6, 400–401.

57. *National Education. A Series of Letters in Defence of the National System against the attacks of an Anonymous Writer in the Sydney Morning Herald by the Teachers of the National Schools at Sydney,* 1857 (copy in Mitchell Library, Sydney), 17.

58. Ibid., 18.

59. Ibid., 18–19.

60. B. Smith, "William Wilkins' Saddle Bags: State Education and Local Control," in Theobald and Selleck, *Family School and State in Australian History.*

61. *A Series of Letters,* 19. By the 1860s, many of the local patrons of national schools were themselves also parents. Mostly drawn from the middle class and comprising tenant farmers, tradesmen, storekeepers, and civil servants, it has been suggested that they were anxious for their children's social mobility. See Jean Ely, *Reality and Rhetoric: An Alternative History of Australian Education* (Sydney: Alternative Publishing, 1978), 12–13 and attached charts.

62. P. Loveday and A. Martin, *Parliament Factions and Parties: The First Thirty Years of Responsible Government in New South Wales 1856–1889* (Melbourne: Melbourne University Press, 1966).

63. Ely, *Reality and Rhetoric.*

64. W. Wilkins, *National Education. An Exposition of the National System of New South Wales; Being the substance of Two Lectures delivered in the Non-vested National School, Pitt Street, Sydney* (Sydney, 1865). Copy in the Mitchell Library, Sydney.

65. Ibid., 6–19.

66. Ibid., 54.

67. Ibid., 54–56.

68. D. C. Griffiths, *Documents on the Establishment of Education in New South Wales 1789–1880* (Melbourne: Australian Council for Educational Research, 1957); Turney, *William Wilkins, His Life and Work* (Sydney: Hale & Iremonger, 1992).

69. Haines, *Lay Catholics and the Education Question.*

70. A. W. Martin, *Henry Parkes A Biography* (Melbourne: Melbourne University Press, 1980).

71. *Public Education Speech of The Hon. Henry Parkes on Moving that the Public Schools Bill be Read a Second Time in the Legislative Assembly, September 12, 1866.* (Copy in Mitchell Library, Sydney), 1–15.

72. Ibid., 21.

73. Ibid., 35.

74. Martin, *Henry Parkes,* 246–248.

75. A. W. Martin, "Henry Parkes: Man and Politician," in *Melbourne Studies in Education 1960–61* (Melbourne: Melbourne University Press, 1962).

76. P. D. Davis, "Bishop Barker and the Decline of Denominationalism," in Turney, *Pioneers of Australian Education.*

77. "Menie Parkes to Henry Parkes, July 1, 1869," in *Letters from Menie: Sir Henry Parkes and His Daughter,* ed. A. W. Martin (Melbourne, Melbourne University Press, 1983), 75.

78. A. R. Crane, "The New South Wales Public Schools League 1874–1879," in *Melbourne Studies in Education 1964* (Melbourne: Melbourne University Press, 1965).

79. D. Morris, "Father J. E. Tenison Woods and Catholic Education in South Australia," in Turney, *Pioneers of Australian Education, vol. 2* (Sydney: Sydney University Press, 1972).

80. "A Pastoral Letter of the Archbishops and Bishops Exercising Jurisdiction in New South Wales, June 1879" in *Select Documents in Australian History Vol. 2 1851–1900,* Manning Clark (Sydney, Angus and Robertson, 1977), 722.

81. D. Morris, "Henry Parkes—Publicist and Legislator," in Turney, *Pioneers of Australian Education,* B. Bridges, "Sir Henry Parkes and the New South Wales Public Instruction Act, 1880," in *Melbourne Studies in Education 1975* (Melbourne: Melbourne University Press, 1975).

82. Martin, *Henry Parkes,* 306.

83. "An Act to Make more Adequate Provision for Public Education, 16 April 1880," reproduced in *Documents on the Establishment of Education in New South Wales 1789–1880,* ed. D. C. Griffiths (Melbourne: Australian Council for Educational Research, 1957), 163–169.

84. Sir Henry Parkes, *Fifty Years in the Making of Australian History* (London: Longmans Green, 1892), 313.

85. "Windeyer, William Charles," *ADB,* vol. 6, 420–422.

86. E. W. Dunlop, "The Public High Schools of New South Wales, 1883–1912," Journal of the Royal Australian Historical Society, 51 (1965). Accessed December 27, 2006 at www.rahs.org.au/jrahs.htm

87. G. Sherington, "Families and State Schooling in the Illwarra," in Theobald and Selleck, *Family, School and State in Australian History;* G. Sherington, "Education and Enlightenment," in *Under New Heavens Cultural Transmission and the Making of Australia,* ed. Neville Meaney (Sydney: Heinemann, 1990).

88. Sherington, "Public Commitment and Private Choice."

89. Parkes, *Fifty Years in the Making of Australian History,* 318–319.

90. Cited in William J. Reese, "Changing Conceptions of Public and Private in American Educational History," in Aldrich, *"Public or Private Education?"* 151–152.

91. Reese, "Changing Conceptions."

92. Martin, *Henry Parkes.*

93. O'Donoghue, *Upholding the Faith.*

94. Brett, Australian Liberals, 2003.

2

Education and State Formation Reconsidered: Chinese School Identity in Postwar Singapore

Ting-Hong Wong

Using the historical case of state regulation of Chinese school identity in Singapore from 1945 to 1965, this chapter demonstrates the dialectical relation between state formation and education. In this chapter, I consider state formation as the historical trajectory through which a governing regime builds or consolidates its dominance. It entails the tasks of cultivating national identity, advancing social and national integration, winning support from the subordinated, and outmaneuvering antagonistic forces.[1] State formation is by no means a unitary project, because the ruling authorities, constantly under pressure from multiple fronts, have to deal with many contradictory demands. There is no guarantee that the ruling regime can handle the conflicting challenges of state formation simultaneously and smoothly.[2] As a result, state hegemonic strategy around schooling, as with state policies in other areas, always brings about contradictory results as far as state formation is concerned. Hegemonic strategies may help the ruling authorities to cope with some crucial challenges of state building, but may, at the same time, leave some important problems of education unresolved. To prevent those unmet demands from endangering their dominant position, the ruling authorities always need to adjust their strategies or bring in auxiliary tactics for additional rounds of intervention. Because of this contradictory nature, state building is an ongoing process of struggle, and the relations between education and state formation are always reciprocal and dynamic.

In this chapter, Chinese schools are considered to be schools using the Chinese language as the chief medium of instruction, and their identity is defined by their cultural exclusiveness. The identity of Chinese schools is most categorical when they are the only institutions in the entire educational system to use the Chinese language as a teaching medium and the teaching of Chinese is the prerogative of Chinese institutions.[3] In the immediate postwar years, Chinese schools in Singapore possessed such an identity. They taught in the Chinese language, and the cultural-linguistic abilities that their students cultivated were totally different from those of their counterparts in other schools. This circumstance hindered the progress of state formation because at that time Singapore was undergoing decolonization, and the state authorities were eager to integrate the Chinese and other ethnic communities into a national whole. To resolve this problem, the British authorities attempted to replace the whole category of Chinese schools with English institutions. However, these policies provoked determined resistance from the Chinese, who were the numerical majority in Singapore. Consequently, the ruling regime was forced to recognize Chinese schools as an integral category within the educational system, deserving treatment equal to those of other institutions. This equalization strategy helped the ruling regime to secure more support from the Chinese people, but it preserved the sharp cultural-linguistic cleavage between Chinese and other schools and hindered the formation of a common national identity. To overcome these problems, the Singapore government looked to dilute the identity of Chinese schools by promoting Chinese learning in English institutions. However, under pressure from the other major racial group, the Malays, they could use this Sinicization approach only minimally. Therefore, Chinese schools preserved a discrete identity, and the compartmentalized system of education continued to produce social fragmentation and to plague state formation.

To explicate this complicated argument about education and the state, I will begin by providing some background about the problem of Chinese schooling in multiracial Singapore.

Background: Chinese Schools in Multiracial Singapore

Singapore, a small island adjoining the southern tip of the Malay Peninsula, is a multiracial society comprising three major local ethnic communities, namely, the Malays, the Chinese, and the Indians. Singapore was founded by the British in 1819. Five years later, an Anglo-Dutch agreement made the Malay Peninsula and Singapore the exclusive preserve of London. In 1826, Singapore was combined with Penang and Malacca, two small territories on the western coast of the peninsula, into one administrative unit that was later known as the Straits Settlements (SS). The British planned to turn the SS into a cosmopolitan trade center.[4]

On the peninsula, between the 1870s and 1910s, the British combined four states into the Federated Malay States (FMS) and another five states into the Unfederated Malay States (UMS) by signing treaties with the Malay sultans of the states concerned. These agreements ensured that the Malay sultans were the ruling partners of the British and upheld Malays as the only indigenous group.[5] The pacts obliged the Malay rulers to act on the advice of the Residents appointed by London on all questions besides those pertaining to Malay customs and religion, with the rulers in the UMS enjoying more autonomy than their FMS counterparts.[6] This pro-Malay constitution was not extended to Singapore, whose sovereignty was ceded in full by the Malay sultans to the British in 1824.[7] However, as the SS, the FMS, and the UMS (together known as British Malaya) were administratively entwined—though the three units had different governing systems—the special relation between the British and the Malays in the much larger Malay Peninsula spilled over to the island and tilted the colonial authorities there toward the Malays.[8]

By the mid-nineteenth century, the Malays enjoyed unquestionable numerical superiority in British Malaya.[9] But with the consolidation of colonial rule, the burgeoning tin and rubber industries attracted a continual flow of immigrants, mainly from China though also from India, and changed the demographic composition of British Malaya. In 1931, the Malays still outnumbered the Chinese in the peninsula. However, if the peninsula and Singapore, where approximately 80percent of the residents were Chinese, were combined, the Chinese population (1,709,392) would exceed that of the Malays (1,644,173).[10] Notwithstanding this change, the Anglo-Malay pacts, which recognized only the position of the Malays, continued to consign the majority of Chinese to the status of aliens.

Chinese residents in Singapore began running their own schools as early as the 1820s.[11] Later, with the expansion of the Chinese community, Chinese schools proliferated. However, the British colonial authorities, bound by their agreements with the Malays, unambiguously proclaimed their educational policy as providing only Malay schools in rural areas and a small number of English schools in the towns.[12] They neither granted a substantial amount of funding to the Chinese schools nor incorporated Chinese institutions as an integral part of the educational system in the colony. This policy left Chinese schools as mainly private institutions that were funded and operated by the local Chinese community and exposed these institutions to external influences from China.[13]

In sharp contrast to the indifferent policy of the British, the Chinese government, conceiving support from Chinese abroad as vital to its campaign for national strengthening, actively used an overseas education policy to maintain political and cultural ties with compatriots in foreign lands. Influenced by educational change in China, modern Chinese schools, founded in China to meet the challenge of the Western powers, replaced *ssushu*, or old-style, Chinese schools that taught Confucian classics in Singapore during the early twentieth century.[14] Beginning in the 1920s, Chinese schools in Singapore registered with the Chinese government, followed the official curriculum of China, and taught in Mandarin—the standard teaching medium mandated by the government of China.[15] Consequently, Chinese schools in Singapore were culturally classified apart from other educational institutions. They inculcated a China-centered outlook in young people, and the linguistic abilities that their students developed were distinct from those of their counterparts in English, Malay, and Tamil schools, which followed the curriculum models in England, Malay Peninsula, and India, respectively.[16]

Besides compartmentalizing the three local racial groups, this divisive educational framework also created a cleavage among the Singaporean Chinese.

By and large, the Chinese people in Singapore were divided into Chinese-speaking and English-speaking Chinese (also called the Straits Chinese or Babas), with the former group being much larger than the latter. Most Chinese-speaking Chinese had either no formal education or had been educated in Chinese schools. Many Babas were the offspring of Chinese-Malay marriages. They grew up in a less Sinicized environment, went to English schools, which admitted pupils from all races, and became deracinated from their Chinese culture and language. Being culturally Anglicized, many Babas became successful professionals; some were appointed by the British to various positions of the colonial state. Because of this background, the two groups of Chinese were culturally and socially segregated,[17] though this too was to change later on. This compartmentalizing nature of the Chinese schools did not concern the colonial authorities too much before World War II, as at that time Singapore was still a colony and the British were under no pressure to blend the "Chinese aliens" and other ethnic groups into a national whole. However, the circumstance changed dramatically after the war.

Most Chinese schools in Singapore were closed during the Japanese occupation between 1942 and 1945, but after the war they resurrected rapidly. In 1946, when school enrollments in Singapore totaled 76,609 students, 46,699 (61percent) of them attended Chinese schools.[18] With the new challenges of state formation in the postwar era, Chinese schools, having a discrete identity, posed serious problems to the ruling regime. First, owing to anticolonial movements in Singapore and the larger international arena, the small island launched decolonization shortly after the war.[19] Therefore, it was imperative for the government to construct a local identity shared by all ethnic communities. Second, the British authorities were under tremendous pressure to improve relations between the Chinese and the Malays, which had deteriorated during the Japanese occupation when the Malays were used by the Japanese as collaborators in carrying out many anti-Chinese measures.[20] In this context, Chinese schools, creating a divisive identity, were considered to block the cultivation of a Singaporean consciousness and interracial harmony.

Furthermore, anti-Chinese mobilizations from the Malays put Chinese schools deeper into conflict with the demands of state formation. Immediately after the war, the British advanced a plan to combine Malacca, Penang, and the nine states in the peninsula into the Malayan Union, leaving Singapore as a separate colony. This scheme, which meant withdrawing many special privileges of the Malays and giving local citizenship to the Chinese in the Union under lenient terms, elicited strong opposition from the Malays. Finally, the British compromised by replacing the Union with a plan for the Federation of Malaya. The new scheme preserved most privileges of the Malays and required the Chinese to fulfill more stringent requirements for citizenship.[21] But after that, the Malays, now perceiving their interests to be in conflict with those of the Chinese, used their position in the Federation to advance many anti-Chinese policies in citizenship, language, and education.[22] Though only about 15percent of the population in Singapore, now a British colony itself, was Malay, the anti-Chinese mobilizations in the peninsula influenced the Chinese school policies of the Singapore government. During that time it was generally believed that Singapore, a small territory with no natural resources, had ultimately to become part of the Federation.[23] Therefore, if Chinese schools continued to promote a Chinese-centered identity and equip their students with cultural-linguistic abilities that were very different from those of their counterparts at other schools, Singapore would be hindered from integrating with the peninsula. These forces prompted the ruling regimes of Singapore to diffuse the cultural peculiarities of Chinese schools.

Notwithstanding these pressures, some contradictory forces restrained the government from tampering with the identity of Chinese schools. In the first place, alarmed by the Malays' onslaughts, Singaporean Chinese swiftly mobilized to safeguard their position in the local setting. Considering Chinese schools as crucial to the preservation of their language and ethnic identity, Chinese people adamantly opposed any state policy that was perceived as strangling Chinese education. The voice of the Chinese residents was impossible to ignore, especially after the mid-1950s, when the majority of them were enfranchised and the system of popular election was installed because decolonization was in progress.[24] Second, should the ruling authorities intervene to weaken the cultural distinctiveness of Chinese schools, they would alienate more Chinese and hand the Malayan Communist Party (MCP), which aimed at toppling the British and all subsequent ruling powers backed by London, an opportunity to enlarge the oppositional movement.[25] It was under these conflicting challenges of state formation in the two postwar decades that the three ruling regimes—the British, the Labour Front (LF), and the People's Action Party (PAP)—tried to resolve the problems engendered by Chinese schools promoting a discrete and Chinese-centered identity.

The Substitution Approach

The first strategy used by the postwar Singapore government to regulate Chinese school identity was substitution. Substitution was the most unyielding form of cultural intervention, for its objective was to replace the whole category of Chinese schools with English institutions. It was adopted between the late 1940s and the early 1950s, when the British still monopolized state power in the colony. This unpopular strategy took shape after several stages of policy formation. It elicited determined opposition from the Chinese masses and failed to bring about the desired results.

The Ten-Year Education Plan and Its Supplementary Program

In August 1946, Arthur Creech Jones, the Parliamentary Under-Secretary of State for the Colonies, requested that Singapore prepare an educational plan for the next five years. Shortly afterward, J. B. Neilson, the Director of Education in Singapore, drafted a paper titled "A Plan for Future Educational Policy in Singapore."[26] Based upon the educational canon that "the first step of education should be through the mother tongue of the child,"[27] Neilson proposed free primary education using the media of Malay, Chinese, Tamil, and English, with English schools being only for pupils whose mother tongue was English.[28]

Christopher Cox, the Education Adviser of the Secretary of the State for Colonies, received the proposal unfavorably. He preferred to see steps taken to desegregate the racially based vernacular schools.[29] Later, in the Advisory Council, an interim body installed by the British Military Administration for consultation, the majority of members endorsed the notion that English schools should accept all students whose parents elected to have their children educated in English. They considered that "English is still the most important language in this country," that the taxpayers had the right to send their children to English schools, and that English schools could promote racial integration.[30]

These opinions left their mark on the final policy. In August 1947, when the Ten-Years' Program was finalized, the government announced that any parents might send their children to English primary schools, the "nursery for the Malayan-minded," if they wished.[31] Although acknowledging the difficulty of correcting the racial nature of vernacular schools, the final plan continued to suggest granting free places to deserving pupils in approved Chinese and Indian schools and expanding the grant-in-aid system for these institutions. After all, with an insufficient number of English schools, the colonial authorities still relied upon vernacular institutions for educational provision.[32]

Developments after the promulgation of the Ten-Years' Program made Chinese schools more recalcitrant toward the colonial authorities and finally prompted the authorities to replace the Chinese schools. With the MCP launching a violent insurrection in 1948 and the escalation of civil war in China, Chinese schools in Malaya were increasingly swayed by the Communists, who aimed to topple the British, and China-oriented forces (such as the Kuomintang and the China Democratic League), which cultivated loyalty toward a foreign land (China).[33] These developments were antithetical to the demands of state formation of the British authorities, who were struggling to defeat the Communists and promote a local consciousness. These events prompted the Commissioner-General's Conference to conclude that the only way to stop the undesirable activities in Chinese schools was to provide more English schools for the Chinese.[34] Almost at the same time, the British authorities were finding it difficult to bring Chinese schools into their orbit by means of financial subsidization. In March 1949, the Education Department proposed increasing financial assistance to Chinese middle schools from S$48,386 to more than S$200,000 on the condition that they modeled themselves after English schools.[35] However, the management of eight middle schools insisted that increased subsidization should be unconditional and rejected the offer.[36] In June 1949, the Singapore representative declared in the Conference of Directors of Education of Southeast Asian British Territories that his government had decided not to grant additional financial support to Chinese schools.[37]

The colonial regime published the *Ten-Years' Program: Data and Interim Proposals* in September 1949. The document stated that "the need of literacy in English in a polyglot Singapore society is overriding" and that if all parents were free to choose, a large number who were now sending their children to vernacular schools would elect to send them to English schools.[38] Two months later, the government put into effect a supplementary scheme, which planned to erect 18 buildings for English schools each year from 1950 to 1954, in addition to the schools planned to be constructed under the original Ten-Years' Program. When all these extra schools were fully used, a policy document stated, English school enrollments would rise from 42,000 to 128,400, while enrollments in vernacular, mainly Chinese, institutions would be drastically reduced from 72,000 to 25,000.[39] The government expected this program to equip the whole Asian population in Singapore with English-speaking ability and a Malayan consciousness.[40]

This substitution policy triggered resistance from the Chinese community. The Singapore Chinese School Teachers Association (SCSTA) queried whether the cultivation of a Malayan consciousness would be possible only in English schools.[41] The MCP condemned the substitution policy as a conspiracy to destroy Chinese culture. To ensure that its schools could survive the impending onslaught, the Hokkien Huay Kuan (the Hokkien Clansmen's Association) campaigned to raise funds for its education activities.[42]

The British took a further step in the substitution scheme when they unveiled the Chinese school subsidization plan. In March 1951, they announced they would increase Chinese school subsidization by 100 percent.[43] However, the British also amended the Education Code and tightened their control over Chinese schools. Statement 31 of the revised code stated that when there were enough places in the lowest grade of English schools to cater to all six-year-old children in the colony, the Director of Education would rescind all the grants-in-aid and remissions of fees for first-year students in Chinese schools. This revocation would be extended to second-year students in the subsequent year and so on, until no Chinese schools received state subsidies.[44] More ominously, the Education Department announced that when the government started withdrawing Chinese school subsidization, English schools would be free.[45] This change revealed that the government eventually wished to oust Chinese schools.

The Failure of the Substitution Approach

The new subsidization policy brought forth more organized opposition from the Chinese community. On June 9, 1951, a conference attended by representatives from more than 200 Chinese schools unanimously resolved to condemn the new scheme. The meeting decided to enlist support from external bodies, including the Singapore Chinese Chamber of Commerce (SCCC)—the overarching Chinese association in the local community—and the Legislative Council, for further petitions to the British authorities.[46] One month later, a letter of protest was drafted by representatives of five schools on behalf of all Chinese institutions in Singapore. Invoking the United Nations Charter, which required all governments to aid vernacular education, the memorandum urged the colonial government to shelve the new subsidization scheme.[47]

Besides the opposition of the Chinese people, the substitution approach also suffered from inadequate school-building capability that resulted from a series of miscalculations. First, the colonial regime had misjudged the growth of the school-age population. In 1949, it was predicted that in 1959 there would be 217,000 children in the 6-to-12 age group.[48] But four years later, the government found that the size of the postwar baby boom had been underestimated and that the correct prediction should be 258,129.[49] Second, the colonial state's school-construction capability was badly affected by material shortages caused by the outbreak of the Korean War.[50] Third, competing claims from other departments and the incompetence of the Public Works Department further limited the government's capacity to provide schools.[51] Because of these pitfalls, the government managed to complete the construction of only 19, 9, and 10 English primary schools in 1951, 1952, and 1953, respectively, all below the annual target of 23 set by the Ten-Years' Program and its Supplementary Program.[52] With such a slow rate of school construction, the British could hardly contract the size of the Chinese school sector, let alone eliminate it. In 1953, of the 160,782 pupils in the colony, 79,272 (49.3 percent) attended Chinese schools, while enrollments in English institutions were 71,003 (44.2 percent).[53] This developmental trend was at variance with the projection of the colonial state that by combining the Ten-Years' Program and its Supplementary Program only about 30 percent of students in the colony would still be in Chinese schools in 1954.[54] In mid-1953, the Deputy Financial Secretary announced to the Education Finance Board that the building programs under the Ten-Year and Supplementary programs would not proceed.[55] Obviously, the colonial government needed another strategy to subdue the Chinese education sector.

The Anglicization Approach

The Anglicization approach blurred the identity of Chinese schools by changing their medium of instruction to English. This hegemonic tactic was a reasonable alternative after the setback of the substitution method. As the colonial government did not have the capability to replace Chinese schools, a better option was to transform the latter into a less threatening form. Policymakers expected that by bringing the teaching medium of Chinese schools in line with that of the English schools, those vernacular institutions would stop producing a divisive identity and create better Singapore citizens. This policy, nevertheless, also elicited strong opposition from the local Chinese.

The Bilingual Education Policy

In October 1953, Governor John Nicoll decided that Singapore's school system, with less than half of its pupils learning entirely in English, did not augur well for building a homogeneous society. To remedy this defect, he proposed a policy of bilingual education, which required Chinese schools to use both English and Chinese as the teaching media.[56] D. McLellan, the Director of Education, later explained that the government would increase subsidization for Chinese schools on the condition that the latter Singaporeanize themselves, which meant teaching in two languages and following a Singapore-centered curriculum.[57] Once again, these proclamations antagonized the Chinese community. The Singapore Chinese School Conference (SCSC) asserted that children should give priority to their mother tongue and that if Chinese schools needed to upgrade their English, English schools should also improve their Chinese.[58] *Sin Poh*, a leftist vernacular newspaper, criticized the suggestion to Singaporeanize Chinese institutions. It averred that the curriculum of English schools, which followed the pedagogic model of England, was also not Singapore-centered.[59]

Despite animosity from the Chinese community, the policy paper "Chinese Schools—Bilingual Education and Increased Aid" was put before the Legislative Council on December 8, 1953. Emphasizing the significance of giving pupils of Chinese schools a "working knowledge of English and Chinese" and turning out good citizens of the colony rather than just good Chinese, the paper proposed to double the financial subsidization of Chinese middle and primary schools that were currently receiving government grants and to create a new subsidization grade for all approved Chinese schools that were getting no state funding at that time.[60] To be eligible for these improved terms, "schools should aim at a curriculum in which the time devoted to the teaching of English and of other subjects in the medium of English would be in the Primary school at least one-third, in the Junior Middle school one-half, and in the Senior Middle School two-thirds of the total teaching time".[61] This bilingual plan would raise the state's expenditure on Chinese school grants-in-aid from the original estimate of S$1,986,000 to S$3,136,600 in 1954.[62]

Resistance and Concession: Keeping Anglicization at Bay

The bilingual education scheme triggered another round of opposition from the Chinese community. After the proposal was passed, the SCSC urged the SCCC and other related bodies to organize joint action.[63] The SCCC asserted that Chinese parents had the right to choose the kind of education they wished for their children. It also accused supporters of the bilingual policy of betraying their Chinese ancestors and offspring.[64] In early January 1954, a joint meeting of representatives from 139 schools and the SCCC unanimously resolved to oppose the bilingual education policy.[65] A 17-person delegation from this meeting then met the Acting Director of Education. Afterward, a joint press release stated that the delegation endorsed the spirit of bilingual education but considered any stipulation on teaching hours using English unnecessary.[66] Leaders from the SCCC and schools then elected seven people to draft a memorandum opposing the bilingual education plan.[67]

In the face of this opposition, McLellan compromised by offering a new interpretation of the original policy paper. He explained that the thrust of the policy was merely to ensure Chinese schools had a good standard in two languages and that the proposed proportion of teaching hours using English was just a guideline. He also assured that the colonial government was not concerned with the number of teaching hours in English as long as the English standards of Chinese school in primary year 6 and junior and senior middle 3 levels were comparable to English school years 3 (primary 3), 5 (primary 5), and 7 (secondary 1), respectively.[68] McLellan invited all Chinese schools to apply for the new grants-in-aid.[69]

The colonial authorities made this concession after realizing that the Anglicization project was jeopardizing other imperatives of state formation, namely, winning support from the Chinese people and outmaneuvering the MCP. Indeed, when they tried to Anglicize the Chinese schools, they alienated the Chinese masses and gave the Communists the opportunity to expand their influence. With growing evidence of MCP infiltration into education and many other spheres beginning in late 1953,[70] the colonial regime's primary concern was now to secure the cooperation of the Chinese community in general and Chinese school authorities in particular for its anti-Communist maneuver. As a result, the British decided to compromise, at least temporarily, on the goal of merging various educational streams. This move, no longer forcing Chinese schools to change their teaching medium, effectively withered resistance—a meeting between the SCCC and the Chinese School Management/Staff Association (CSM/SA) resolved to advocate for Chinese schools to join the new grants-in-aid program.[71] A week before the deadline on March 31, 1954, 78 schools had applied for new subsidization.[72]

Several important events in the following year propelled the government further away from the strategy of Anglicization. First, Singapore changed its institutional rules following its transformation from a colony into a self-governing state. In April 1955, Singapore held its first significant public election. Afterward, a predominantly popularly elected legislative assembly was formed, and the Executive Council was replaced by a ministerial cabinet, with six of its nine members coming from the government in power. These changes brought the state and civil society closer and introduced new state actors whose outlooks were very different from those of the colonial bureaucrats. More important, after introducing the automatic registration of voters, the legitimacy of the state depended increasingly upon the consent of Chinese voters, who now became the majority among the electorates.[73] Against this context, all political forces contending for state power had to reckon with the interests of the Chinese. Second, a string of class strikes and riots in 1955 suggested growing Communist influences among students of Chinese schools.[74] Badly in need of the school authorities' cooperation in its anti-Communist maneuvers, the government had to avoid suggesting any policy that might alienate the Chinese. Finally, the struggle for equal treatment of Chinese education peaked in the mid-1950s. On June 6, 1955, a conference of delegates from 503 Chinese associations, ranging from the most conservative to the most radical, formed the Chinese Education Committee (CEC) to fight for equal treatment of the Chinese language and schools.[75] This conference was larger than all previous similar gatherings. Later, evidence revealed that the MCP had captured this campaign, as the leftists secured many influential positions within the CEC. Facing a unified movement capitalized on by its chief antagonist, the popularly elected government refrained from tampering with the medium of instruction in Chinese schools.

The Equalization Approach

Equal treatment of Chinese education was first championed by the Singapore state as an official policy in the 1956 Education White Paper. This policy, upholding Chinese schools as an integral category of the educational system in Singapore, in effect preserved the identity of the Chinese institutions. The policy of equalization was formulated after the LF government, the first popularly elected regime in Singapore, accepted the recommendations of an all-party committee appointed in the midst of a crisis caused by an industrial conflict. Unlike the policies of substitution and Anglicization, the equalization tactic was devised after the popularly elected politicians listened to people's representations and articulated the interests of the Chinese masses.

Escalating Social Tension and the All-Party Committee on Chinese Education

In late April 1955, an industrial conflict with the Hock Lee Bus Company worsened. Large numbers of students from Chinese middle schools arrived day after day to support the strikers. After a series of futile mediations, tension escalated and the support to the strikers, mainly from students and other labor unions, grew swiftly. On May 12, hostility culminated in riots that killed 4 persons and injured 31.[76] Blaming the students, the government decided to close the three largest Chinese middle schools—the Chinese High and the two Chung Cheng schools.[77] Immediately, students in these three institutions converged to protest the government's decision.[78] Students from other Chinese middle schools swiftly issued a public statement reprimanding the government.[79] On May 17, the LF government offered to reopen the three schools on the condition that the latter expel a number of "ringleaders" and show cause why they should not be declared unlawful. This announcement further provoked the students. The same evening, some 2,000 students gathered in Chung Cheng and barricaded themselves inside the school buildings. The next day, students from five other middle schools staged a strike. This protest was also buttressed by many labor unions.[80]

Because of this disturbance, other political parties in the Legislative Assembly vehemently attacked the LF. The LF government then appointed a committee, whose nine members were all from the political parties holding seats in the assembly, to investigate the situation in Chinese schools.[81] This step was probably taken because the LF, as a governing party lacking mass support, wished to enhance its legitimacy by offering increased participation.[82] The committee successfully convinced the government to reopen the three schools unconditionally.[83] It then invited public opinion on Chinese education.

The *Report of the All-Party Committee of the Singapore Legislative Assembly on Chinese Education,* which information came from memoranda from 87 individuals and associations and interviews with 14 people, was put before the Legislative Assembly in early February 1956. Seeking a clean break with the previous substitution and Anglicization strategies, the report postulated that suppressing the language of the Chinese, who made up more than 80 percent of the population in Singapore, was not practical. It suggested endorsing the position of Chinese education so that Chinese culture, together with the cultures of other ethnic groups, could contribute to a nation marching rapidly toward independence.[84] Based upon these premises, the committee recommended melding Chinese schools into the education system of Singapore by giving them equal treatment.[85] It advocated extending the full grants-in-aid already given to English-aided schools to Chinese institutions and paying teachers of Chinese schools on the same terms as their counterparts in English schools.[86] The report also upheld Mandarin as the medium of instruction in Chinese schools.[87] In effect, this policy protected the identity of Chinese institutions.

Mindful of the fact that a polyglot society such as Singapore needed some "languages of wide communications" as lingua francas, the committee proposed that Chinese schools teach English as the second language and Malay or Tamil as the third. The *All-Party Report* recommended increasing public expenditure on Chinese schools to more than S$15 million per year. In return, the school authorities were expected to follow the state-promulgated Singaporeanized curriculum, reform their management committees, improve discipline, and keep students away from party politics and industrial disputes.[88] The LF government adopted most of these recommendations in its March 1956 White Paper on Education Policy.[89]

The Equalization Approach and Splitting the Oppositional Movement

Eliciting two different types of reactions from the Chinese community, the equalization approach served state formation by splitting the social movement of Chinese education and isolating the MCP, the chief adversary of the British and the LF. After the release of the *All-Party Report,* conventional Chinese bodies such as the SCSC and the Singapore Chinese Middle School Teacher Association responded only mildly, as their primary concerns were the protection of Chinese culture and the material conditions of Chinese schools.[90] However, the leftist bodies reacted vehemently. For instance, a meeting of some 40 Chinese school old boys' associations condemned the government for using conditional subsidization to interfere with the internal operation of Chinese institutions.[91] The Singapore Chinese Primary School Teacher Association (SCPSTA) accused the committee of scapegoating Chinese schools for the social unrest and attempting to impose the language of a retreating imperial power.[92] Almost all these criticisms were echoed by the Singapore Chinese Middle School Student's Union, another leftist body.[93]

When the White Paper (released in late March 1956) finalized the equalization policy, the schism in the Chinese community was even more conspicuous. While moderate bodies such as the CSM/SA and the SCSTA did not strongly oppose the reform plan,[94] radical associations such as the SCPSTA, the CEC, the Chinese school old boys' associations, the Teachers of English in Chinese School Association, and the Pan-Malayan Students' Union, fervently protested the White Paper.[95] More importantly, the leftists openly criticized the lukewarm attitude of the SCCC in opposing the policy. The SCCC's president was, according to the constitution passed by 503 Chinese associations on June 6, 1955, the ex officio chairman of the CEC.[96] In late April 1956, the SCCC unilaterally withdrew from the CEC,[97] probably because it had became clear that the Communists already controlled the committee. After all, as the LF had promised to give Chinese schools equal treatment and retain Mandarin as their medium of instruction, many Chinese people felt that the state policy had became more acceptable.

The equalization approach served state formation by isolating the leftists and winning more Chinese people's support for the state. In 1959, Singapore held another general election. Now the Chinese residents, almost all of whom had been enfranchised by the Citizenship Ordinance passed in 1957, had even more power to determine the fates of all contending forces.[98] Therefore, it was not surprising that the PAP, which ultimately won the election, pledged to support the policy of equal treatment for Chinese schools when electioneering.[99] Nevertheless, the equalization policy had one serious shortcoming as far as state formation was concerned. By leaving the compartmentalized system of schooling generally intact, it failed to bridge the cultural-linguistic gap between people educated in Chinese and other schools and hindered the formation of a common Singapore identity. To overcome this weakness, the ruling authorities needed to find another policy to complement the equalization approach. Sinicization was such an auxiliary strategy.

The Sinicization Approach

Beginning in the mid-1950s, the ruling regimes in Singapore—first the LF and then the PAP—adopted Sinicization to supplement the tactic of equalization. Through this policy, the government attempted to weaken the identity of Chinese schools by promoting the teaching of Chinese in English institutions. Without interfering directly with Chinese schools, this approach was unlikely to antagonize the Chinese masses. The LF and the PAP expected the Sinicization strategy to advance state formation by narrowing the cultural and linguistic cleavage between people educated in Chinese and those educated in other schools. In addition, if implemented successfully, this policy would no longer allow the teaching of Chinese to be the prerogative of Chinese institutions and would reduce the importance of Chinese schools in preserving the Chinese language and culture. As a result, it might preempt the Chinese people from reacting furiously, should an occasion arise and the government need to replace Chinese institutions in the future. Nevertheless, the Sinicization strategy was unsatisfactorily executed because English schools in Singapore had a weak legacy of Chinese teaching and, more importantly, the LF and the PAP were hamstrung by the Malays.

Sinicization of English Schools under the British

For several reasons, the colonial government in Singapore had put minimal effort into introducing the teaching of Chinese in English schools in the prewar period. In the first place, because of its geographical location, Singapore had never been considered a crucial stepping-stone to penetrate China, especially after Hong Kong became a British colony in the 1840s. Thus the colonial government and the Christian bodies—the major sponsors of English schools in Singapore—were not strongly motivated to teach Chinese.[100] Second, given the special relations between the British and the Malays, the colonial government was concerned more with accommodating the Malays, rather than the Chinese, in educational policy.[101] As a result, English schools in prewar Singapore provided some teaching of Malay but not Chinese; Chinese culture remained outside the curriculum of English institutions.[102] Because of this background, the British had to start from a weak foundation when they wished to employ the strategy of Sinicization after World War II.

The first postwar years witnessed increased efforts by English schools in Singapore to promote Chinese. On August 12, 1946, a Chinese newspaper reported that two English institutions, Raffles and Victoria, had started offering Chinese classes to prepare students for the Chinese paper of the Cambridge Certificate Examination—a test for students completing secondary education in English schools.[103] The following year, R. M. Young, the Acting Director of Education, disclosed that since the end of the war the government had been experimenting with teaching of Chinese language in English schools.[104] Nevertheless, these Sinicization moves failed to dilute the cultural distinctiveness of Chinese schools. First, the trial schools had difficulties hiring suitable instructors. Because of the a limited number of Chinese classes offered, the school authorities preferred teachers of Chinese to be bilingual so that they could also teach other subjects in English. Second, the government did not pay teachers of Chinese, whose positions did not exist in the approved establishment of English schools. Hence, many experimental schools were compelled to maintain these "special classes" by charging extra fees or soliciting donations from outside sources.[105] These constraints reined in the Sinicization of English institutions. In 1947, only 129 English school students sat for the Chinese paper in the Cambridge Certificate Examination; in 1948, the number was 291.[106] Worse still, the standard of the Chinese language taught in English schools was quite low: a Chinese newspaper reported that the Chinese paper in the Cambridge Certificate Examination required candidates to have knowledge of Chinese only equivalent to that of the Primary 5 level in Chinese institutions.[107]

The Sinicization movement received another stimulus in the early 1950s, when the Ten-Year Supplementary Plan, a scheme to substitute Chinese schools by expanding English institutions, was adopted. To rebut the Chinese populace's accusation that the colonial regime was obliterating Chinese culture, E. C. S. Adkins, the Secretary of Chinese Affairs, advocated serious efforts to introduce the teaching of Chinese in the new English schools.[108] Afterward, the Education Department decided that from September 1951, all 51 government English primary schools would provide Chinese classes for Chinese pupils.[109] A. W. Frisby, the Director of Education, also revealed that vernacular language classes in 36 government schools built under the Supplementary Plan were already free, and aided English schools might start providing free vernacular language classes the following year.[110] When the new school term started in September 1951, the government posted 47 newly hired teachers of Chinese to English schools.[111] In 1952, R. Watson-Hyatt, the Chief Supervisor of Chinese Schools, announced that about 20 new teachers of Chinese would be hired in English primary institutions.[112] In the following year, the government planned to hire another 37 such teachers.[113]

Despite these moves, the progress made by the Singapore colonial government in adding Chinese to mainstream institutions was not very impressive. In the mid-1950s, Chinese was still merely an optional subject in English primary schools. As for English secondary schools, the government still had no policy to promote Chinese at that level. In early 1954, an investigation by the Department of Education found that only 1,807 students in government and aided secondary schools were studying Chinese, Malay, or Tamil. Even among schools teaching vernacular languages, these subjects, often offered outside regular school hours, were not part of the ordinary school curriculum.[114] Furthermore, since at that time the colonial government's policy was to pay for teachers of vernacular languages only in primary schools, many pupils taking Chinese in secondary institutions had to pay extra fees.[115]

Colonial officers' apprehension about the political loyalties of teachers of Chinese might also have limited the Sinicization plan. In 1952, Watson-Hyatt divulged that before employing a teacher of Chinese, the government would ask the Special Branch to scan his or her political record. This procedure was meant to ensure that the "evil of Communism" or "Chinese chauvinism" would not contaminate English schools.[116] With such unimpressive progress with Sinicization, the Singapore colonial state could hardly convince the local Chinese that their culture was being preserved in English schools, and the LF and PAP regimes faced a big challenge to institutionalize the teaching of Chinese language in English institutions.

Sinicization: A Subsidiary Tactic of the Equalization Strategy

In the mid-1950s, when the LF attempted to regulate the identity of Chinese schools through equalization, the Sinicization tactic assumed a new role in state formation. Given that the government had promised to grant all four kinds of schools equal treatment and to preserve Mandarin as the teaching medium of Chinese institutions, the sector of Chinese education was bound to exist for many years to come. To bridge the cultural-linguistic gap between people educated in different streams of schools, the *All-Party Report* suggested improving the standards of teaching Chinese in English schools.[117] It also advocated that "Mandarin should be the only language to be taught for all Chinese pupils as the compulsory second language in English schools."[118]

The LF swiftly inaugurated a committee to improve the Chinese syllabus and teaching method in English schools. It also expanded the squad of instructors of Chinese for English institutions.[119] In 1957, the LF decided that all government and aided English schools should offer their students at least one period of vernacular language per day. However, a shortage of suitable staff forced the

Ministry of Education to compromise on this decision "slightly."[120] The Sinicization plan was also hampered by the fact that the study of Chinese language was, by and large, inconsequential to pupils in English primary schools—until the late 1950s, Chinese was not made a subject for the common entrance examination for English secondary schools.[121]

In 1959, the PAP won the general election and became the new ruling power in Singapore. The new challenges of state formation placed the ruling power in a deeper contradiction as far as the Sinicization scheme was concerned. On the one hand, like the previous LF government, the PAP had to secure support from the Chinese masses and rebut the accusation that they were destroying Chinese culture. Because of these concerns, the PAP government pledged to continue the policy of giving equal treatment to Chinese schools and safeguarding Chinese culture.[122] This position restrained the PAP from changing the medium of instruction of Chinese schools and forced Prime Minister Lee Kuan Yew and his associates to find another solution to weaken the cultural barrier between Chinese schools and other institutions.

One of the solutions identified by the PAP was the Sinicization of English institutions. When the PAP first came to office in 1959, there were 336 teachers of Chinese working in 134 English institutions, but by 1963, there were 605 such instructors in 155 English schools. In 1960, the PAP made the second language an optional subject for the Primary School Leaving Examinations. Two years later, the second language was counted as half a unit, of a total of five, in the same exam.[123] In 1963, the weighting was increased to one, out of a total of six units.[124]

Nevertheless, other challenges of state formation reined in the Sinicization practices of the PAP. After coming to office, Lee Kuan Yew and his associates actively pursued complete independence from London by a merger with the Malay Peninsula, where the Alliance government was dominated by the United Malays National Organizations (UMNO). This undertaking led to two additional challenges of state formation, namely, to integrate Singapore and the peninsula into a national whole and to win the Malays' consent to PAP dominance in the island. These demands suppressed the introduction of Chinese teaching in English schools, because one of the major conditions for merger was to ensure that the educational policy in Singapore was approximate with that in the peninsula, where only Malay and English were mandated as compulsory subjects in schools. Besides, to integrate Singapore with the peninsula, Lee Kuan Yew wished to narrow the linguistic differences between the two territories. Consequently, the PAP actively promoted Malay, the dominant language in the peninsula.[125] This additional language as a school subject competed with Chinese for resources and timetable space in English schools. It also, in effect, relegated Chinese to the position of third language. Furthermore, to allay the suspicion of the Malays, who regarded Singapore as a Chinese chauvinist stronghold threatening the predominance of Malay in the peninsula, the PAP had to prevent being labeled as pro-Chinese.[126]

Perhaps because of these factors, the PAP required only candidates who had at least three years of continual instruction in a language other than the teaching medium of their schools to sit for the second language paper in the Primary School Leaving Examination in 1963, the year Singapore joined the Federation of Malaysia.[127] In effect, this rule allowed pupils from English schools to skip learning Chinese. Also, among candidates from English schools sitting for the second language examination, a substantial proportion did not opt for the Chinese paper; for example, in 1963, 27 percent of such candidates sat for Malay, 5 percent for Tamil, and 68 percent for Chinese.[128]

The turbulence that Singapore went through after joining Malaysia further curbed the PAP from promoting Chinese teaching in English schools. Being part of Malaysia, the PAP now faced more pressure to avoid being tainted as pro-Chinese, for that meant anti-Malay from many Malay people's point of view. In 1964, racial tension escalated when the PAP made a mistake by entering nine candidates for the Malaysian general election.[129] This injudicious move contravened the "gentlemen's agreement" between Lee Kuan Yew and Tunku Abdulah Rahman, the leader of the UMNO, that the UMNO and the PAP would refrain from contesting in each other's territories.[130] Worse, it also exacerbated the Malays' fear that the PAP, whose leaders and members were predominantly Chinese,[131] aspired to challenge Malay supremacy in the peninsula.[132] Malay extremists reacted by campaigning against the "dictatorial Chinese PAP government led by Lee Kuan Yew." This animosity finally deteriorated in July and September into two violent racial clashes, which claimed 36 lives. During both these riots, the Singapore government imposed a curfew in the island.[133]

This anti-Chinese hostility made the PAP cautious about initiating further Sinicization moves. In late 1964, the Ministry of Education announced that second language would be made a compulsory subject in all schools the following year. With this change, the teaching of Chinese would be offered in more English schools.[134] However, this plan did not materialize, and second language was still only an optional subject in English secondary schools in late 1965.[135] Even after Singapore's withdrawal from Malaysia in August 1965, the PAP continued to be hamstrung in promoting Chinese language and culture because the small island, now a tiny and vulnerable nation, had to avoid provoking its neighboring Muslim countries—Malaysia and Indonesia.[136]

As the English schools had been Sinicized only to some extent, Chinese schools, whose identity remained discrete, continued to hinder state formation by producing people with identity and cultural-linguistic traits at variance with those educated in other schools—even decades after Singapore had gained independence.[137] Furthermore, with the limited Sinicization of English institutions, many Chinese residents in Singapore continued to regard Chinese schools as indispensable to preserving the "root" of Chinese culture and language.[138] Many Chinese people, especially the Chinese-educated, were infuriated after seeing a series of PAP policies culminate in the "extinction" of Chinese schools in 1987. This lingering sense of frustration and anger could be easily exploited by the oppositional forces for anti-PAP agitation, as demonstrated by the case of Tang Liang Hong in the 1997 Singapore general election.[139]

Conclusion

In this chapter I have discussed the dialectical relations between state formation and education. State formation, as stated in the introduction, is not a coherent project, for it contains the multiple and conflicting tasks of cultivating national identity, advancing national or social integration, winning support from the subordinated, and outmaneuvering antagonistic forces. Since the ruling authorities are constantly under pressure on multiple fronts, there is no guarantee that they can meet these challenges of state building simultaneously and smoothly. Because of their contradictory natures, state hegemonic strategies related to schooling always have conflicting ramifications for state formation—they may help the ruling regime cope with some crucial challenges of state building, yet at the same time leave other problems unresolved. To prevent those unmet demands from imperiling their dominant position, the ruling authorities need to modify their strategies for future rounds of intervention. This chapter has used the historical case of state regulation on Chinese school identity in postwar Singapore to concretize this theoretical claim.

Immediately after Word War II, Singapore launched decolonization. During that time, Chinese schools had a discrete cultural identity. They promoted a sectional, Chinese-centered outlook and equipped students with cultural-linguistic abilities that were very different from those of their counterparts at other schools. Therefore, they prevented the British from accomplishing two significant tasks of state formation, namely, creating a common, Singapore-centered identity and promoting interracial integration. The colonial regime sought to resolve this problem by eliminating the whole category of Chinese education through large-scale expansion of English schools and by imposing English as the teaching medium in Chinese institutions. However, these tactics provoked adamant resistance from the Chinese, and the MCP exploited the situation to enlarge its oppositional campaign. In other words, by employing these strategies, the British were unable to handle another crucial challenge of state formation – namely, to win consent from the subordinated and to outmaneuver the antagonists.

Since the mid-1950s, when the majority of Chinese residents became enfranchised, the two popularly elected governments—the LF and the PAP—were forced to adjust their strategy and to uphold Chinese schools as an integral and distinct category in the Singapore education system deserving equal governmental support. This new approach served state formation because it won support from some Chinese people and curtailed the following of the MCP. However, by keeping the compartmentalized school system intact, it compromised the imperatives of promoting a common national identity and cultivating integration among the Chinese and other ethnic groups—other crucial tasks in state building.

To minimize the disintegrative effects unleashed by a divisive education system, both the LF and the PAP endeavored to blunt the cultural distinctiveness of Chinese schools through the supplementary tactic of strengthening Chinese teaching in English schools. Nevertheless, this Sinicization strategy was only minimally implemented, chiefly because the ruling regimes in Singapore were constrained by their relations with the Malays. The Malay factor became especially prominent when Lee Kuan Yew actively pursued merger with the Federation of Malaya. To promote smoother national integration with the peninsula, the PAP had to rein in the promotion of Chinese learning in English schools lest they be attacked as Chinese chauvinists and the relations with Kuala Lumpur became strained. As Lee Kuan Yew and his associates had only minimally Sinicized English schools, the cultural identity of the Chinese institutions remained strong, and the education system continued to hamper state formation by maintaining a sharp social cleavage—even decades after Singapore had gained independence.

This historical case of Singapore bears some important theoretical implications and prompts us to rethink the connections between education and state formation. It reminds us that state formation is a complicated project containing multiple and conflicting tasks and that state authorities, constantly under pressure from diverse fronts, are always faced with dilemmas when exercising hegemonic strategies in the educational sphere. These insights have been overlooked by many scholars studying education and the state.[140] Through the complicated story of state regulation of Chinese schools in Singapore, this chapter prompts us to remember the contradictory consequences of state educational policies as well as the dialectical connection between education and state formation.

Abbreviations

Declassified Confidential Files
    CO (Colonial Office) 537/7288
    CO 717/162/52746
    CO 825/90/7
    CO 1030/426
SCA (Secretary for Chinese Affairs, Singapore) 10/1953
    SCA 25/1951
    SCA 69/54
    Newspapers
    Nan Chiau Jit Poh (NCJP)
    Sin Chew Jit Poh (SCJP)
    Sin Poh (SP)
    Straits Budget (SB)

Notes

1. This definition is mainly inspired by Andy Green. See Green, *Education and State Formation: The Rise of Education Systems in England, France, and the USA* (New York: St. Martin's Press, 1990); Andy Green, "Technical Education and State Formation in Nineteenth-Century England and France," *History of Education* 24 (June 1995): 123–139.

2. Michael W. Apple, "Critical Introduction: Ideology and State in Education Policy," in *The State and Education Policy,* Roger Dale (Philadelphia: Open University Press,

1989), 1–20; Michael W. Apple, *Education and Power,* 2nd ed. (New York: Routledge, 1995); Martin Carnoy and Henry M. Levin, *Schooling and Work in the Democratic State* (Stanford, CA: Stanford University Press, 1985); Roger Dale, *The State and Education Policy* (Bristol, PA: Open University Press, 1989); Claus Offe, *Contradictions of the Welfare State* (Cambridge, MA: MIT Press, 1984); Miles Ogborn, "Law and Discipline in Nineteenth Century English State Formation: The Contagious Diseases Acts of 1864, 1866 and 1869," *Journal of Historical Sociology* 6 (March 1993): 28–55; Derek Sayer, "Everyday Forms of State Formation: Some Dissident Remarks on 'Hegemony,'" in *Everyday Forms of State Formation: Revolution and the Negotiation of Rule in Modern Mexico,* ed. Gilbert M. Joseph and Daniel Nugent (Durham and London: Duke University Press, 1994), 367–377; Ting-Hong Wong, *Hegemonies Compared: State Formation and Chinese School Politics in Postwar Singapore and Hong Kong* (New York: Routledge, 2002).

3. This idea of cultural exclusiveness is inspired by Basil Bernstein's notion of classification, which is used to analyze boundary maintenance among different subjects in school curriculum. See Bernstein, *Class, Codes and Control,* vol.3 (London: Routledge, 1975); *Class, Codes and Control,* vol. 4, *The Structuring of Pedagogic Discourse* (London: Routledge, 1990). Although this paper defines the identity of Chinese schools by their degree of exclusiveness in using Chinese as teaching medium and offering subjects in Chinese studies, it does not mean that the identity of these institutions is purely determined by these dimensions. I realize that other aspects—such as the content of school knowledge—are also crucial in molding the identity of Chinese schools. See Wong, *Hegemonies Compared.*

4. Stanley S. Bedlington, *Malaysia and Singapore: The Building of New States* (Ithaca and London: Cornell University Press, 1978), 31; Yeo Kim Wah, *Political Development in Singapore, 1945–55* (Singapore: Singapore University Press, 1973), 1.

5. Albert Lau, *The Malayan Union Controversy: 1942–1948* (Singapore: Oxford University Press, 1990), 3–19.

6. Bedlington, *Malaysia and Singapore,* 33–34.

7. Edwin Lee, *The British as Rulers: Governing Multiracial Singapore, 1867–1914* (Singapore: National University of Singapore, 1991), 3–19.

8. James de Vere Allen, "Malayan Civil Service, 1874–1941: Colonial Bureaucracy/ Malayan Elite," *Comparative Studies in Society and History* 12 (1970): 149–187; Yeo Kim Wah, *Political Development in Singapore,* 1–13, 69.

9. Khoo Kay Kim, "Sino-Malaya Relations in Peninsular Malaysia before 1942," *Journal of Southeast Asian Studies* 12 (March 1981): 93; Victor Purcell, *The Chinese in Southeast Asia* (London: Oxford University Press, 1965), 234.

10. C. A. Vlieland, *British Malaya: A Report on the 1931 Census and on Certain Problems of Vital Statistics* (London: Waterlow, 1932).

11. Tan Liok Ee, *The Politics of Chinese Education in Malaya, 1945–1961* (Kuala Lumpur: Oxford University Press, 1997), 8.

12. H. R. Cheeseman, "Education in Malaya 1900–1941," *Malaysia in History* 22 (May 1979): 126–137; Philip Loh Fook Seng, *Seeds of Separatism: Educational Policy in Malaya, 1874–1940* (Kuala Lumpur: Oxford University Press, 1975); Puteh Mohamed and Malik Munip, "The Development of National Educational System," *Malaysia in History* 28 (1985): 76–93.

13. Saravanan Gopinathan, *Towards a National System of Education in Singapore, 1945–73* (Singapore: Singapore University Press, 1974); Tan Liok Ee, *Politics of Chinese Education;* Harold E. Wilson, *Social Engineering in Singapore: Educational Policies and Social Change, 1819–1972* (Singapore: Singapore University Press, 1978).

14. Lee Ting Hui, "Chinese Education in Malaya, 1894–191 —Nationalism in the First Chinese Schools," in *The 1911 Revolution—The Chinese in British and Dutch Southeast Asia*, ed. Lee Lai To (Singapore: Heinemann Asia, 1987), 48–65; Wong, *Hegemonies Compared,* 50–51.

15. Gopinathan, *Towards a National System,* 4–5.

16. Gopinathan, *Towards a National System;* Loh, *Seeds of Separatism;* Keith Watson, "Rulers and the Ruled: Racial Perceptions, Curriculum, and Schooling in Colonial Malaya and Singapore," in *The Imperial Curriculum: Racial Images and Education in the British Colonial Experience,* ed. J. A. Mangan (London and New York: Routledge, 1993), 147–174; Wilson, *Social Engineering in Singapore.*

17. Owing to this cultural gap, the English-educated Chinese were generally not sympathetic to the struggles of the Chinese-speaking Chinese for Chinese language and education in postwar years. See Jurgen Rudolph, *Reconstructing Identities: A Social History of the Babas in Singapore* (Aldershot: Ashgate Publishing, 1998); Yong Ching Fatt, "A Preliminary Study of Chinese Leadership in Singapore, 1900–1941," *Journal of Southeast Asia History* 9 (September 1968): 258–285.

18. Wilson, *Social Engineering in Singapore,* 143.

19. Nicholas Tarling, *The Fall of Imperial Britain in South-East Asia* (Singapore: Oxford University Press, 1993).

20. Cheah Boon Kheng, "Sino-Malay Conflicts in Malaya, 1945–1946: Communist Vendetta and Islamic Resistance," *Journal of Southeast Asian Studies* 12 (March 1981): 108–117; Cheah Boon Kheng, *Red Star over Malaya: Resistance and Social Conflict during and after the Japanese Occupation, 1941–1946,* 2nd ed. (Singapore: Singapore University Press, 1983).

21. James de Vere Allen, *The Malayan Union,* Monograph Series No. 10 (Yale University, Southeast Asia Studies, 1967); Cheah Boon Kheng, "Malayan Chinese and the Citizenship Issue, 1945–48," *Review of Indonesia and Malayan Affairs* 12, no. 2 (December 1978), 95–122; Albert Lau, *A Moment of Anguish: Singapore in Malaysia and the Politics of Disengagement* (Singapore: Times Academic Press, 1998); Mohamed Noordin Sopiee, *From Malayan Union to Singapore Separation: Political Unification in the Malaysia Region, 1945–65* (Kuala Lumpur: Penerbit Universiti Malaya, 1974); Ting-Hong Wong and Michael W. Apple, "State Formation and Pedagogic Reform in Singapore, 1945–1965," *Comparative Education Review* 46, (May 2002).

22. Donna Amoroso, "Dangerous Politics and the Malay Nationalist Movement, 1945–47," *South East Asia Research* 6 (November 1998): 253–280; Yeo Kim Wah, *Political Development in Singapore.*

23. Lau, *A Moment of Anguish;* Yeo Kim Wah, *Political Development in Singapore;* Michael Hill and Lian Kwen Fee, *The Politics of Nation Building and Citizenship in Singapore* (London: Routledge, 1995).

24. Wong, *Hegemonies Compared;* Yeo Kim Wah, *Political Development in Singapore;* Wilson, *Social Engineering in Singapore.*

25. Purcell, *The Chinese in Southeast Asia;* Watson, "Rulers and the Ruled"; Wilson, *Social Engineering in Singapore.*

26. Gimson to Creech Jones, January 11, 1947, CO 717/162/52746.

27. "Educational Policy in the Colony of Singapore: The Ten-Years' Program," adopted in Advisory Council on August 7, 1947 [hereafter cited as "The Ten-Years' Program"], Appendix III, 23.

28. "The Ten-Years' Program," Appendix I, 11.

29. Comments of the Educational Advisor, included in Paskin to Gent, April 8, 1947, CO 717/162/52746.

30. "The Ten-Years' Program," Appendix IV, 30–33.
31. "The Ten-Years' Program," 2; 7.
32. Ibid., 6–9.
33. Richard Clutterbuck, *Conflict and Violence in Singapore and Malaysia* (Singapore: Graham Brash, 1984); Kwei-Chiang Chiu, *Changing National Identity of Malayan Chinese, 1945–59* [in Chinese] (Xiamen, Fujian: Xiamen University Press, 1989); Wong, *Hegemonies Compared.*
34. "Minutes, the Ninth Commissioner-General's Conference", January 22 and 23, 1949, CO 717/162/52746. This conference was organized by the Office of Commissioner-General in Southeast Asia, a regional body installed by the Whitehall to coordinate anti-Communist activities in the British territories of that area.
35. *SCJP,* February 3, 1949; *SB,* March 3, 1949.
36. *SCJP,* February 19, 1949.
37. "Report, Conference of Directors and Deputy Directors of Education," June 23 and 24, 1949, CO 717/162/52746.
38. *Ten-Year Program: Data and Interim Proposals.* (Department of Education, Colony of Singapore, September 1949), 1–2.
39. *Supplement to the Ten-Year Program, Data and Interim Proposals* (Department of Education, Colony of Singapore, undated) [hereafter cited as *Supplement Program*], 118–119.
40. *SB,* February 22, 1950.
41. *NCJP,* July 28, 1950.
42. Translation, *Freedom News,* issue 23, June 15, 1951, CO 537/7288; Appendix B, Report on a Vernacular Publications Bureau, June 13, 1951, CO 825/90/7.
43. *SCJP,* March 4, 1951.
44. *SCJP,* May 12, 1951.
45. *SCJP,* May 16, 1951.
46. *SCJP,* June 10, 1951.
47. *SCJP,* July 4, 1951.
48. *Supplement Program,* 118.
49. *Proceedings of the Second Legislative Council* [hereafter cited as *PSLC*], Colony of Singapore, 1st Session, 1951 and 3rd Session, 1953, October 20, B320.
50. *PSLC,* 1st Session, 1951, October 16, B296–298.
51. *PSLC,* 3rd Session, 1953, October 20, B322.
52. Wong, *Hegemonies Compared.*
53. Wilson, *Social Engineering in Singapore,* 143.
54. *Supplement Program,* 119.
55. "Minutes, Education Finance Board," June 19 and 20, 1953, SCA 10/1953.
56. *SB,* October 22, 1953.
57. *SP,* November 7, 1953.
58. *SCJP,* November 27, 1953.
59. *SP,* November 7, 1953.
60. "Chinese Schools—Bilingual Education and Increased Aid," in *PSLC,* 3rd Session, 1953, Colony of Singapore, No. 81 of 1953 [hereafter cited as "Bilingual Education"], col. 542–545.
61. Ibid., col. 547.
62. Ibid., col. 544.
63. *SCJP,* December 19, 1953.
64. *SCJP,* January 5, 1954.
65. *SCJP,* January 6, 1954.

66. *SCJP,* January 12, 1954.
67. *SCJP,* January 17, 1954.
68. *SCJP,* March 4, 1954.
69. *SCJP,* March 7, 1954.
70. Clutterbuck, *Conflict and Violence;* Lee Ting Hui, *The Open United Front: The Communist Struggle in Singapore, 1945–1966* (Singapore: South Seas Society, 1996).
71. *SCJP,* March 4, 1954.
72. *SCJP,* March 23, 1954.
73. Yeo Kim Wah, *Political Development in Singapore;* Yeo Kim Wah and Albert Lau, "From Colonialism to Independence, 1945–1965," in *A History of Singapore,* ed. Ernest C.T. Chew and Edwin Lee (Singapore: Oxford University Press), 117–153; Lau, *A Moment of Anguish.*
74. Clutterbuck, *Conflict and Violence;* Lee Ting Hui, *The Open United Front.*
75. *SCJP,* June 7, 1955 and *SP,* June 7, 1955.
76. Clutterbuck, *Conflict and Violence,* 108–109; *SCJP,* May 13, 1955; *SB,* May 19, 1955.
77. *SB,* May 19, 1955.
78. *SCJP,* May 15, 1955.
79. *SP,* May 17 and 19, 1955.
80. *SCJP,* May 21, 1955; *SP,* May 18, 19, and 21, 1955.
81. *Report of the All-Party Committee of the Singapore Legislative Assembly on Chinese Education* (Singapore: Government Printer, 1956) [hereafter cited as *All-Party Report*], 1.
82. In the 1955 General Election, the LF won merely 26 percent of the votes and 10 positions out of the 25 total contested seats. They became the majority in the Assembly only because the fledging PAP entered only four candidates, and old powers such as the Progressive and the Democratic parties, led by English-speaking elites, had no strong Chinese-speaking followings. See Yeo Kim Wah and Albert Lau, "From Colonialism to Independence."
83. *SCJP,* May 22, 1955.
84. *The All-Party Report,* 4.
85. Ibid., 7.
86. Ibid., 19, 34.
87. Ibid., 42.
88. Ibid., 9–10, 39–44, 15–27.
89. "White Paper on Education Policy," Legislative Assembly, Singapore, Sessional Paper, No. Cmd. 15 of 1956 [hereafter cited as "White Paper"], 5.
90. *SCJP,* March 1 and 6, 1956.
91. *SCJP,* March 5, 1956.
92. *SCJP,* March 6 and 7, 1956; and *SP,* March 8, 1956.
93. *SCJP,* March 19, 20, and 22, 1956.
94. *SCJP,* April 2 and 15, 1956.
95. *SCJP,* April 4, 11, and 12, 1956; *SP,* April 3, 1956.
96. *SP,* April 10, 1956.
97. *SP,* April 28, 1956.
98. Ong Chit Chung, "The 1959 Singapore General Election," *Journal of Southeast Asian Studies* 6 (March 1975): 61–86.
99. *SB,* April 12, 1959.
100. Wong, *Hegemonies Compared.*
101. Khasnor Johan, *The Emergence of the Modern Malay Administrative Elite* (Singapore: Oxford University Press,1984); Rex Stevenson, *Cultivators and Administrators: British*

Educational Policy toward the Malays, 1875–1906 (Kuala Lumpur: Oxford University Press, 1975); Yeo Kim Wah, "The Grooming of the Elite: Malay Administrators in the Federated Malay States, 1903–1941," *Journal of Southeast Asian Studies* 11 (September 1980): 287–319.

102. *Suggestive Course of Instruction and Syllabus for English Schools in the Straits Settlements and Federated Malay States* (Kuala Lumpur: Government Printer, 1939); Wong, *Hegemonies Compared,* 150.

103. *SCJP,* August 12, 1946.

104. *SCJP,* April 25, 1947.

105. Ibid.

106. *SCJP,* March 19, 1949.

107. *SCJP,* August 12, 1946.

108. Appendix B, Report on a Vernacular Publications Bureau, June 13, 1951, CO 825/90/7.

109. *SCJP,* July 2, 1951.

110. *SB,* July 13, 1951.

111. Minutes, Singapore Education Committee, September 26, 1951, SCA 25/1951.

112. *SCJP,* February 12, 1952.

113. *Annual Report, Department of Education* [hereafter cited as *ARDE*], Colony of Singapore, 1953, 46.

114. *ARDE,* 1953, 50.

115. Minutes, Singapore Education Committee, March 31, 1954, SCA 69/54.

116. *SCJP,* February 14, 1952.

117. *All-Party Report,* 17.

118. Ibid., 40–41.

119. *SCJP,* June 11, 1956 and December 1, 1957.

120. *First Education Triennial Survey, Colony of Singapore, 1955–57* (Singapore: Government Printer, 1959) [hereafter cited as *FETS*], 31.

121. Appendix A, Minutes, Conference of Directors of Education held at Government Office, Brunei Town, State of Brunei, October 13–15, 1958, CO 1030/426.

122. Gopinathan, *Towards a National System,* 34; Wilson, *Social Engineering in Singapore.*

123. *SCJP,* January 29, 1963.

124. Ministry of Education, Annual Report, 1963 [hereafter cited as MEAR], 10.

125. Gopinathan, *Towards a National System,* 33–37.

126. *Petir Weekly* (in Chinese), no.1 (July 18, 1959): 6–8.

127. MEAR, 1963, 10.

128. Ibid., 12.

129. Yeo Kim Wah and Albert Lau, "From Colonialism to Independence," 144–145.

130. Michael D. Barr, "Lee Kuan Yew in Malaysia: A Reappraisal of Lee Kuan Yew's Role in the Separation of Singapore from Malaysia," *Asian Studies Review* 21 (July 1997): 1–17; Lau, *A Moment of Anguish;* Sopiee, *From Malayan Union.*

131. Pang Chen Lian, *Singapore's People's Action Party: Its History, Organization and Leadership* (Singapore: Oxford University Press, 1971), 36.

132. However, it is important to note that the leaders of the PAP were predominantly English-speaking Chinese, especially after the leftist leaders broke away to form another party—the Barisan Socialis—in 1961.

133. Lau, *A Moment of Anguish.*

134. *SCJP,* November 7, and December 24, 1964.

135. *SCJP,* November 14, 1965.

136. Bedlington, *Malaysia and Singapore;* Chan Heng Chee, The Politics of Survival, 1965—1967 (Singapore: Oxford University Press, 1971).

137. Kwok Kian Woon, "Social Transformation and Social Coherence in Singapore," *Asiatiche Studien Etudes Asiatiques* 49, no. 1 (1995): 217–241; Sai Siew Yee, "Post-Independence Educational Change, Identity and Huaxiaosheng Intellectuals in Singapore: A Case Study of Chinese Language," *Southeast Asian Journal of Social Science* 25, no. 2 (1997): 79–101.

138. Wong, *Hegemonies Compared.*

139. Tang, a Chinese-educated lawyer without a previous record of political activism, was a candidate of the Workers' Party in the 1997 election. He campaigned mainly by accusing the PAP of discriminating against Chinese education and won many voters' support, though he failed to win. To intimidate opponents trying to imitate the same strategy in future campaigns, the PAP adopted all means possible to harass Tang after the election. Tang finally fled Singapore. See James Chin, "Anti-Christian Chinese Chauvinists and HDB Upgrades: The 1997 Singapore General Election," *South East Asia Research* 5 (November 1997): 217–241; Wong, *Hegemonies Compared.*

140. John Boli, *New Citizens for a New Society: The Institutional Origins of Mass Schooling in Sweden* (Elmsford, NY: Pergamon Press, 1989); Bruce Curtis, *True Government by Choice Men? Inspection, Education, and State Formation in Canada West* (Toronto: University of Toronto Press, 1992); Andy Green, *Education and State Formation;* Andy Green, "Technical Education and State Formation," 123–139; Andy Green, *Education, Globalization and the Nation State* (London: Macmillan Press, 1997); Stephen L. Harp, *Learning to be Loyal: Primary Schooling in Nation Building in Alsace and Lorraine, 1850–1940* (Dekalb: Northern Illinois University Press, 1998); and James Van Horn Melton, *Absolutism and the Eighteenth-Century Origins of Compulsory Schooling in Prussia and Austria* (New York: Cambridge University Press, 1988).

3

How the State Made and Unmade Education in the Raj, 1800–1919

Tim Allender

This chapter examines colonial education efforts in India between 1800 and 1919, particularly the imposition and consequences of systemic state-directed schooling. The complexity of the subcontinent precludes a grand narrative approach. But the chapter argues that as the nineteenth century progressed, the state's increasingly active hand in education was a primary reason for its eventual disengagement from the broader Indian population. This happened well before *swadeshi* (self-sufficiency, in opposition to British colonial governance) was to become an effective force against British-led education in the twentieth century.

Before 1858, governance in colonial India centered on the rule of the East India Company, and after that time, it emanated directly from the India Office in London via the Viceroy's council in Calcutta and the various local governments in each province and presidency. However, education in India had many involutions in the nineteenth century, and developments could be highly regionalized. Before elaborating on these, the following summary may be a useful starting point.

Unlike England, Australia, and North America, colonial rulers in India had to navigate complex language barriers and preexisting political and social orders. The agency of the individual who experimented in the field was also a significant factor in developing educational thought, even though there were relatively few European personnel running schools and colleges before 1854. Both these factors, and a strong indigenous educational heritage, provided a rich context for highly imaginative and spontaneous "orientalist" thinking to develop in the 1820s and 1830s. This approach offered felicitous possibilities where a confluence of Western and Eastern knowledge might build a sympathetic state schooling system uniquely adapted to the needs of both village and urban children.

However, in the 1830s, work toward this goal began to unravel as the state took a more active role. A new generation of bureaucrats, often imbued with Benthamite utilitarian values imported from England, began to challenge the efficacy of the orientalist approach. As a result of the Anglicist/Orientalist controversy of 1835 Governor-General William Bentinck issued a resolution that decreed that teaching in state-funded schools should be in English rather than in the local languages of each region. A generation later, Charles Wood's ambitious Education Dispatch of 1854 signaled a new India Office policy in support of state education of "the masses." This was after the earlier policy of "filtration," which relied on Western-educated indigenous elites to pass on their education to the general population, had failed. In the 1860s and 1870s, the combined effects of Wood's Dispatch and the Revolt of 1857 saw most state-schooling efforts eventually centralized in the large cities, where newly developed European schooling approaches were artificially imposed. The administrative strategy of "decentralisation" adopted across all Raj departments in 1871 accentuated this artificiality.

By the early 1870s, earlier assumptions by Raj administrators that state schooling would eventually reach the general population were seriously questioned by Indian Civil Service (ICS) officials. And the mission schools and colleges, with their evangelizing agendas, often found themselves in conflict with state education directives. In 1882, the exhaustive Hunter Education Commission provincial hearings illustrated how detached state-run education had become from the needs of the general population even though government statistics told a story of school expansion and better translations of Western knowledge. Lord Curzon's education conference of 1901 and the centralization of education administration in 1910 attempted to remedy the deficiencies brought about by earlier state intervention. However, by this time, the forces of resistance had begun to appropriate much of the educational discourse generated by state-schooling enterprises of the previous century. This helped to undermine the very rationale of the state in British India, and British-led education in India effectively ended when the Indian Education Service (IES) was "Indianised" in 1919.

Early Orientalist Interactions with the State, 1795–1835

At the beginning of the nineteenth century, there was a distinct difference between England and colonial India in the way the state connected with education. In England the story was still one of disconnection. The first translation of Rousseau's *Emile* (1762), and later, uncertain adaptations of Rousseau's ideas, such as Thomas Day's influential *Sandford and Merton* (1783–1789), had set radical intellectual circles alight about the prospect of preserving the perfect nature of the child by carefully controlling his or her education and environment.[1] The writings of Joseph Priestley, Mary Wollstonecraft, and experimenters such as Robert Owen, which developed Rousseau's ideas further, were contested by the traditional public school- and university classics-dominated curriculum that was still determined to beat the "sin" out of the innately wicked schoolboy. But the pre-1832 Reform Act Westminster politicians played hardly any role in this debate.

However, in India, state alignment with education was more deliberate and was part of a broader strategy of reconciliation driven by Raj insecurities about its power base and the need to better understand the sophisticated preexisting social and political orders resident on the subcontinent. This imperative resulted in the intervention of the state in East India Company affairs with a succession of amendments to the younger Pitt's 1874 India Act, especially at the time of the renewal of the Company's charter in 1793, 1813, 1833, and again in 1853. There was also the legacy of Warren Hastings's official policy of "orientalism," which fused this perceptive strategy with a genuine fascination for Eastern learning.

Part of the strategy involved mastering the complex information order belonging to pre-British India.[2] Clever men such as Ram Mohan Roy and Ram Camul Sen helped the British orient their intellectual endeavors and begin a state-schooling enterprise. Mohan Roy, the brilliant leader of the Hindu reform movement in Bengal, studied the theologies of five religions, including Hinduism and Christianity, in their source languages. As the founder of several secondary schools, he supported the assimilation of Western and Eastern knowledge.[3] Camul Sen, an upper-caste Hindu, also collaborated with Europeans in working toward the introduction of Western education in Bengal.[4] The sympathetic environment created by the state and its East India Company then empowered orientalists such as H. H. Wilson, H. T. Prinsep, and J. C. Sutherland to recognize the intellectual integrity and communal significance of education traditions that were a cohesive force in indigenous communities. State support was also given to institutions such as the College of Fort William and the Calcutta Madrassa (college for higher Islamic learning), which became centers of orientalist research and teaching for both Europeans and the local intellectual elite in India.

The orientalist creed in these early years centered on the question of language and the transmission of knowledge rather than schooling for the young in the government-funded classroom. The creed was also fractured and contested partly because Indian literacy was still relatively low and individual orientalists engaged with an indigenous domain that was regionally and communally segmented. But state governance, via the Company, was sufficiently plural and unselfconscious in this early period to encourage educational experimentation among its European agents on the spot. This could still shape, in turn, the way the state approached education in the Raj even though the emphasis was on the transmission of ideas rather than the building of schools.

The official connection between education and the state was also established with the Charter Act of 1813, which embraced state-sponsored education in India for the first time by devoting Rs100,000 to the "improvement of literature, and the encouragement of the learned natives of India . . . and that any schools, public lectures, or other institutions, for the purposes aforesaid . . . shall be governed by . . . the said Governor-general in Council . . . provided that all appointments to offices in such schools, lectureships and other institutions, shall be made by . . . the governments with which the same shall be situated."[5] In passing this act, the British parliament required the East India Company to take responsibility for public education in India at least 20 years before the British government would do the same in Britian.[6] The establishment of the General Committee of Public Instruction (GCPI) in 1823 confirmed a growing state commitment to finding a genuine educational nexus between East and West. This included the reorganization of the Calcutta Madrassa and the Sanskrit College at Benares, the establishment of two Oriental Colleges at Agra and Delhi, as well as providing funding for the printing of Sanskrit and Arabic books and the employment of orientalist scholars to translate Western knowledge into Persian, Sanskrit, and Arabic. State-sponsored inquiries in the presidencies of Madras and Bombay in this decade also revealed 12,498 and 1,705 indigenous schools, respectively. The difference in numbers was partly due to the different European definitions of what constituted a "school," but these figures confirmed a vital village schooling base with which the state could hope to engage in the future.

The esoteric nature of Eastern learning, which was partly built on language, and the dispersed nature of its "native" custodians also encouraged the state to permit its agents the discretion they needed to seek out and shape this knowledge. William Adam's extensive village school surveys were initially carried out in this cooperative spirit. Unlike the earlier surveys for Madras and Bombay, Adam's three reports were a careful digest of categories of school according to religion, gender, and village. He also ventured a relatively high literacy rate of 6.1 percent for males and 3.1 percent for females as a result of the work of these indigenous schools. Under Mohan Roy's influence, Adam's educational surveys were predicated on the belief that the state could be persuaded to redirect public money to engage "the great masses."[7] Adam believed that the long-standing eagerness of the broader population for education in the local tongues justified large-scale state engagement with traditional indigenous schools that had long been part of village communities. This was preferable to just encouraging the translation of Sanskrit and Arabic literature in the few government-funded schools that taught and translated in these languages. As well, the methodology for Adam's Bengal surveys were couched in the following terms to Governor-General William Bentinck: "With the aid of my Pandit [Hindu teacher] and Moulavee [Muslim cleric] and by friendly communication with the respectable inhabitants and learned men of [Bengal I propose to] make an enumeration or list of the various institutions for the promotion of education; classify them according to the denominations of which they may consist, whether Hindoos, Mahomedans or Christians; public private, charitable; examine each institution . . . the nature and extent of the course of instruction in science and learning."[8] And as far as the state's role in education in the future was concerned, Adam asserted: "[T]o know what the country needs to be done for it by Government, we must first know what the county has done or is doing for itself."[9]

Furthermore, Adam's thorough methodologies in surveying indigenous village schools in Bengal and Bihar up to 1838 were powerful exemplars because they illustrated the vibrancy and variety of the thousands of indigenous schools that existed throughout India.[10] The patronage offered by high company officials, namely Thomas Munro, Mountstuart Elphinstone, and John Malcolm, who encouraged native education societies to be set up in Madras and Bombay, was another example of state intersection with education in this early period. These societies aimed at encouraging learning by compiling dictionaries and publishing cross-translated Eastern and Western scholarship.[11] This was in an attempt to seek universal truths and values by sympathetically rendering centuries of layered indigenous heritage.

Finally, there was a quirky but powerful voice concerning Eurasian children that the state had to navigate in this early period. The best example was Andrew Bell and his monitorial school at Madras. The school was designed to corral the illegitimate children of soldiers and to prevent the unsavory display of the vagrant "orphan" of European blood on the streets of the Raj. It merged citizenship with schooling, as these "orphans" were required to work in the government printery to pay for their keep. Scholars have made much of this experiment and the controversy over Joseph Lancaster's similar model in England. Monitorial schools were already in existence in the United States and Europe. However, for Bell, the Indian provenance and its powerful context of state-sanctioned schooling experimentation legitimated his claim to innovation. That the youngest children learned literacy via the sandbox, thus imitating Indian indigenous teaching methods, was also enough to incline Jeremy Bentham to invoke his "Psammographic principle" and to support the pedagogic credentials of Bell's "Madras system" of pupil monitors.[12]

The Imposition of the State-Sponsored English Classroom, 1835–1849

The period of sympathetic and unselfconscious state engagement with indigenous learning and Eastern knowledge was interrupted by the Anglicist/Orientalist controversy of 1835. This happened at the behest of a few key individuals. Two years earlier, H. H. Wilson, a leading orientalist of the age, retired and was replaced as secretary of the GCPI by Charles Trevelyan. Trevelyan was a member of the Church Mission Society and he believed in the merit of Christianizing India. At the India Office in London, James Mill (father of J. S. Mill), a follower of Jeremy Bentham's Utilitarian creed, also opposed the orientalist approach. Mill believed that attempting to rejuvenate Hindu civilization was pointless and he urged the introduction of Western ideas and institutions instead—Trevelyan's and Mill's ideas were complex—Trevelyan because he also supported scholarship in the classical languages of Arabic, Persian and Sanskrit,[13] and Mill because he was a radical in other arenas and supported a reforming curriculum of new science and political economy in Europe.[14] However, their views, among others, coming from quite different standpoints, redirected state conceptions of education in India toward supporting teaching Western knowledge in the English language. With the matter mostly decided by the newly constituted GCPI, Trevelyan's future brother-in-law, T. B. Macaulay, law member of the governor-general's council, issued his lengthy Minute of 1835 elaborating a new rationale for the change.[15] The governor-general, William Bentinck, then issued a definitive resolution in favor of "the promotion of European literature and science among the natives of India [and that] all funds appropriated for the purpose of education would be best employed on English education alone."[16]

The immediate practical application of the repositioning came when state funds were diverted to establish English-instruction schools. This occurred first in the Bengal and Agra presidencies, and the short-term results were promising. The move was popular among Indians desiring a Western education and proficiency in English because they wished to better participate in the commerce of the Raj and to fill government jobs at the higher pay levels offered by the British. Even before 1835, the likelihood of success in attracting enrollments of this kind had been indicated to the GCPI because strategic minorities, including the *bhadralok* (commercial middle class) of Bengal and the Parsis of Bombay, were already pursuing Western knowledge taught in English as a means to greater prosperity. However, the formal change in government policy induced wealthy Indians to make donations to the cause, which helped to establish a strong base for English schools, especially across Bengal. By the mid-1840s, the state had expanded its program of building its own schools, and the number of pupils attending them numbered 5,570 in Bengal and 10,616 in Bombay.[17]

The Anglicist members of the GCPI, and their allies in the government, assumed this was the natural scheme of things because of the "superiority" of Western knowledge. They were less inclined to attribute the popularity of English instruction schools to shorter-term and unsustainable pecuniary incentives available to Indians who wished to work within the ambit of the Raj. But the GCPI, now under the influence of the Anglicist reformers, also supported a more contrived strategy of downward "filtration," according to which Western knowledge would be eventually passed down to the general population by select groupings who were to be first educated in these new schools. The powerful new symbol of state "imposition," as represented by Macaulay's Minute and Bentinck's resolution, cut directly across the strongly held orientalist belief of the need to "engraft" Western knowledge onto indigenous stock and not just to superimpose it. The imposition of the Western-constructed "classroom," and the teaching of English within it, also began to unwind the earlier subtle and informal education alignment between the orientalists and the Company.

In 1835, the state had intervened to direct that instruction in English be carried out in the schools that it funded. But it was the less-publicized clauses of Macaulay's Minute concerning the local languages that most sharply defined the rift with those who supported the orientalist approach in India. On this issue Macaulay asserted: "[T]he dialects commonly spoken among the natives of this part of India contain neither literary nor scientific information, and are moreover so poor and rude that, until they are enriched from some other quarter, it will not be easy to translate any valuable work into them. It seems to be admitted on all sides, that the intellectual improvement of those classes of the people who have the means of pursuing higher studies can at present be effected only by means of some language not vernacular amongst them."[18]

This part of Macaulay's Minute was a direct rebuff to William Adam's approach of supporting state engagement with lower-order village education in the local languages. Macaulay's assertion that local languages were not sophisticated enough to convey Western knowledge helped give orientalist thought greater cogency in opposition to new state policy. Now, more as part of an agreed community of thinkers, these orientalists, and a new generation of neo-orientalists, pursued their ideal of education, taught in the languages of the subcontinent at the "lower" village level instead.

They knew that an authentic transmission of knowledge across the Eastern/Western divide was only possible via connections such as those already made with Brahmin priests, munshis, and other traditional custodians of education who engaged the highly localized communities of India. For example, Lancelot Wilkinson's[19] experiments in Bhopal (an important location for both high Brahmin learning and a place of strong Muslim educational patronage) engaged Sehore pandits to combine traditional Eastern learning with that of the West. Earlier orientalist notions of "engraftment" were verified by his success in conflating the ideas of Newton and Copernicus with Hindu learning traditions and medieval Sanskritic astronomy.[20] The work of Wilkinson and that of Brian Hodgson,[21] in collaboration with local intellectuals, demonstrated that indigenous teachers who taught in local languages had already integrated new Western scientific theories into their teaching of the young and that the arrival of the British had not been a defining moment in the intellectual enrichment of the subcontinent.[22]

With these deepening orientalist understandings, the academic merit of the English instruction apparatus artificially set up by the state began to be questioned. Was it legitimate to teach European history and geography in preference to that of the subcontinent? Could English ever be really taught to junior classes not imbued with its cultural context? By the 1840s, as a result of these concerns and to escape the Calcutta "cantonment" and its English-based teaching, orientalist work moved away from Bengal to the regional centers of the North Western Provinces (NWP), Bombay, and Madras. And in Madras and Bombay at least, a dual policy of education in the local languages for elementary schooling and English teaching in the central schools began to be pursued.

Village Experimentation and the Dispatch of 1854

The tension between the state and the orientalist educators in India soon proved to be unsustainable. There were only a limited number of government jobs that could be filled by English-educated Indians, and most Indians did not have direct dealings with Raj commerce. This meant that state education could not closely connect with the daily needs of the general population if it did not engage with village education in the first instance using the dominant local language of each region. In the late 1830s, Bentinck's successor as governor-general, Lord Auckland, was able to manufacture a compromise by ensuring that the government continued to fund existing Oriental Colleges, such as the Calcutta Sanskrit College and the Calcutta Madrassa, and to provide scholarships for students wishing to study at these institutions. More significantly, the state also abandoned its policy of filtration. This was because of the findings of crucial village school experiments that were carried out in north India in the late 1840s and early 1850s. The findings were to finally persuade state officials in London of the need to focus on providing education for the general population rather than just on the top-down English-instruction approach.

These important village school experiments were carried out in the NWP and, later, in the Punjab. They demonstrated that thousands of language- and religion-specific indigenous schools and teachers across the subcontinent would be marginalized, along with the irreplaceable indigenous educational heritage that they represented, if state policy was not changed.

Understanding this, as early as 1844, Governor-General Henry Hardinge had already sanctioned the foundation of almost 100 schools in Bengal, with a curriculum of "vernacular reading, writing, arithmetic, geography and history of India and Bengal."[23] But, in the later 1840s, it was Wilkinson's ideas that were responsible for convincing Henry Reid (director of public instruction) and the reformer James Thomason (lieutenant governor of the NWP) to begin surveying indigenous schools in eight districts in the NWP with a view to setting up government-funded village schools and even to start closing minor English-instruction schools already established there.[24] An elaborate scheme called "Halkabandi" (circle of villages) was then cleverly worked out by an alliance of orientalists and junior officers answerable to the province's secretariat.

The Halkabandi experiment in the NWP, and later in the Punjab, was on a large scale that covered 2,000 village and *tahsil* (subdistrict) schools teaching in the local languages and that employed teachers formally engaged and trained in the vast array of indigenous schools around the old Mughal centers of Lahore, Delhi, and Agra and their periphery. These experiments represented a high-water mark in state schooling in British India. They were driven by a strong belief that a meeting and crossover of indigenous schooling traditions with their government counterparts could be made more seamless in the future by a sensitive state appropriation of the pandit and the *maulvi* and the rich indigenous learning traditions that they represented.

The schemes were substantial because of the secure funding base that they were given. This was provided by a 1 percent loading on the land tax revenues in each province.[25] As a measure of the relative size of the revenue raised, Reid, in the NWP, calculated that once allocated, it approximated to Rs 4 per month per child per week. This was the equivalent of the three-shilling rate only the most ambitious education reformers hoped for as a funding base when William Forster presented his Education Act in England two decades later, in 1870.[26]

Thomason fed back the detailed findings of the NWP experiment to the India Office. He pressed the case for well-grounded local-language instruction that also delayed teaching in both the court language of Urdu and in English. This, it was argued, would make it possible for at least some poor village children to transcend the language divide and progress up the schooling hierarchy to college and university. The strategy was also forward looking enough to prevent college education from becoming the preserve of the wealthy, as it had become in England. With this in mind, William Arnold argued:

> On the one hand we do not want a College far above the heads of the people, on the other hand still less do we want a mere English school on a large and expensive scale. Our design rather is that as an Education of the humblest kind is

to be afforded to all the children . . . so by means of a Central College an Education of the highest order may be given to those [students] from ability or station [who] are qualified to receive it. A high education must necessarily be reserved for the few, and those few will be composed of two classes. 1st those who are fortunate enough to be able to pay for the luxury of a refined education & 2nd those who by talent and industry have earned a right to be helped . . . While however, we should require the rich to be to a certain extent learned, we ought not require the learned to be rich.[27]

It is true that by this time, growing state acceptance of its role in directing and funding education was influenced by more formalized and broad-based schooling developments in Europe. But these village experiments in north India, spurred on by the orientalist reaction to Macaulay's Minute, brought the India Office to accept a role that was far greater than anything the state had attempted in England by this time. This was because such schemes, and the elevated level of state commitment that accompanied them, were necessary if indigenous schooling was to be sympathetically built upon.

Partly as a consequence of these findings, Charles Wood issued his pivotal Education Dispatch of 1854. The dispatch signaled a more deliberate government attempt to bring education to what it now deemed as "the masses": a phrase that reflected the newly developed "science" of political arithmetic in England.[28] It was also a domestic political document that needed to satisfy important stakeholders (including the missionaries) who had a voice in Westminster. As a result, it superficially embraced elements of both sides of the Anglicist/Orientalist controversy of 1835. It supported the cause of instruction in English at senior schooling levels and it denigrated the "grave errors" of Eastern learning. It also held up as worthy Sanskrit, Arabic, and Persian literature and the role of the oriental scholar in inculcating ethics and "more advanced science." Furthermore, it promoted some elements of orientalist thinking. General education in India, the dispatch asserted, could only be achieved by "a careful attention to the study of the vernacular language of the district . . . [to] be gradually enriched by translations of European books or by original compositions of men whose minds have been imbued by the spirit of European advancement."[29]

State Overgovernance: The Withering Away of the Oriental Connection, 1860–1875

Administrators considered Wood's dispatch the lodestar for education in India. It gave rise to an impressive bureaucratic regulatory model of systemic state schooling in each province. This included a powerful integration of existing government educational efforts and a more formulaic commitment by the state to institutional education across the Raj. The dispatch introduced grant-in-aid regulation that mostly applied to mission schools and a formula for linking village school, district school, college, and university education by scholarships, school inspection and building grants.[30] The dispatch also foreshadowed the founding of three universities, with examination powers reaching those colleges already established in India. It established unitary education departments in each province that institutionalized and connected school inspection and teacher training via the normal school as well as a hierarchy of government village, tahsil, *zillah* (district), and Anglo-vernacular schools that fed into the college.

Unfortunately, Wood's official sanction for a pressing out to the *mofussil* (periphery), to educate "the masses," was given just before the shock of the Revolt of 1857. In conflict with Wood's intentions, the Revolt suddenly prompted frightened ICS officers and their superiors in Calcutta to retreat from all village-based enterprises. This was even though the newly established and provincially based education departments had begun to chart an independent course. They had their own education service personnel, many of whom wanted to further the work of the oriental experimenters of the previous generation. Unlike the ICS, the education departments' recruitment was by an informal network of advice and patronage, and only after 1859 were they required to pass an examination in one Indian language spoken in the region where they were to serve. However, in post-Revolt India, and after the dissolution of the Company, the actions of these education officers were placed under greater scrutiny by the ICS. This was important because the ICS did not share the academic interest of many education officers in "wasteful" experimentation nor were they strongly influenced by the orientalist work and thought of the previous decades. Instead, the ICS ethos was predicated on the reforms of the Northcote-Trevelyan report (1854) that built upon the earlier neutral and technocratic administrative approach of the Utilitarians in India.[31] Entry into the ICS now was by competitive examination, and their belief was in "efficient" and codified administrative procedure.

As a result, ICS officers were not sympathetic to the building of a specialized understanding of indigenous schooling, especially after the breach of trust, as they saw it, represented by the Revolt of 1857. Subservient education officers were now judged by the state using uniform and codified standards of "efficiency" that focused upon enrollment numbers and classroom-building regulations rather than on pedagogy and language sensitivity. The new alignment between the state and its ICS agents in the field also detached the arcane knowledge of the regional educator, and this resulted in the marginalization of the thousands of language- and religion-specific indigenous schools in the 1850s and 1860s: schools that collectively represented a deep and irreplaceable indigenous educational heritage. Liberal thought, and ICS overgovernance, now disrupted the more sensitive orientalist discourse of the 1820s and 1830s.[32] As Bayly asserts, in the more general terms of the period, while the expatriate society in India became more hierarchical and government more a matter of routine, the earlier deeper social knowledge withered away.[33]

The Revolt shifted government attention away from the village to the more easily controlled urban centers, especially in north India. Paradoxically, the more active hand of the state, brought about by Wood's dispatch, also saw a burgeoning in the number of state-run schools in these centers. The establishment of three universities in 1857 gave immediate impetus for the foundation of new colleges. In 1857 there were 27 colleges, and by 1882, their number had grown to 72. Five of these colleges were aided institutions run by Indians, but with European principals, because Indians were considered unfit to hold these positions. Secondary schools increased in number from 169 in 1855 to 1,363 in 1882.

Wood's dispatch had initially planned for indigenous schools to be incorporated into the state's efforts to teach at the primary level. But after the Revolt, a second dispatch issued by Wood's successor, Lord Stanley, in 1859, reversed the strategy, directing that only government schools should be relied upon to spread education to the general population. At this schooling level, each province and presidency was left to develop its own primary education approach, and this ranged from building upon the Halkabandi system for both girls and boys in the NWP to a system of aided primary schooling institutions in Bengal. Bombay, on the other hand, developed primary education directly through its own state schools. Government tallies of these schools were notoriously unreliable as each provincial department attempted to justify its claims of incremental "progress" and "efficiency" in each of its annual reports in the 1860s and 1870s. For example, in 1882, the number of primary schools was put at an inflated figure of 14,486 in Madras and 47,402 in Bombay, while only 6,712 schools were listed for NWP.[34] However, in this latter province, arguably the most effective connection with schooling needs of the village poor was made in this period.

Education was now also subject to tightly controlled protocols of government reporting. The IES and its mission allies numbered over 300 Europeans in the middle of the nineteenth century. After 1860 each provincial government was required to publish an annual education report, which had a formal ICS readership of 400. There were also many other stakeholders both in the Raj and in Great Britain who read these reports. This dwarfed Kay Shuttleworth's Education Department in England (not established until 1839) whose operatives numbered just 50 people in the middle of the nineteenth century.[35] The new state apparatus, with its more systematic level of communication and accountability, was to pave the way for the imposition of another form of innovation, innovation that this time was directly imported from Europe. It was poorly adapted to the education needs of the subcontinent and it would be a powerful force for disengagement with the general population in the coming two decades.

The State Appropriates Western Education for India, 1870–1882

By the late 1860s the state had become an unintentional agency for a more permanent disconnection from the broader Indian population despite its many new schools. Prescient indigenous observers such as Radhakant Deb had anticipated the phenomenon as early as 1851.[36] He saw the new instrumentalist approach of most state-sponsored educational efforts as creating a dangerous void between the state and the general population, unprecedented even in pre-British times. The dynamic of Western experimentation in the more easily controlled urban centers was advantageous to those elites who had become beneficiaries of Raj commerce and patronage. Thirty years later, in provinces such as the Punjab, the census of 1891 confirmed this, showing that the three principal trading castes—the Banyas, the Khatris, and the Aroras—made up 40 percent of the literate population of the province.[37]

However, standardized knowledge disseminated via textbooks cut off local communities from their earlier tentative connectedness with Western intellectuals. In 1865, all provincial education departments wrote holistic, bifurcated curriculums that incorporated English as well as the pre-British court language in each major province. They were driven by Calcutta University's preference for examinations in English, and even at the lower levels of government schooling, simplified but fragmented learning about Western subjects attempted to prepare students for these later examinations.

Attempts to unify European-oriented knowledge, without the crucial local context, saw textbook committees imposed across the Raj in 1873 and again in 1877. Expert translations of Western knowledge into at least the court languages, most notably Urdu, were embarked upon. However, this work was restricted by the difficulty of employing indigenous scholars, who often preferred to remain outside the British ambit and who rejected the British approach as to what should be taught to Indian children. Inappropriate attempts were now also made by the state to impose wholly Western-constructed education strategies, including Payment by Results, Pupil Teachers, and the Middle School, via the powerful bureaucratic structures set up earlier as a result of Wood's dispatch. The pivotal role of the teacher, long esteemed by traditional cultures on the subcontinent, was often ignored as a potential agency for adapting knowledge to better suit the needs of Indian schoolchildren. Government inspectors usually could not speak the local languages. Instead, they relied on using rote learning in English to gauge scholastic success in the schools that they visited. The seeking of alms by the local pandit in the village to supplement his income and to indicate his traditional local status, also offended ICS sensibilities that preferred instead the payment of an inadequate "salary" of Rs 10 to Rs 15 per month for his service.

Particularly in south India, the Christian missions made significant inroads into translating and teaching in this period. Female education and female teacher training flourished in several mission centers, stimulated by the influence of the Unitarian Mary Carpenter and others at the highest levels of government in Calcutta. Unfortunately, the state was far less successful in educating girls. An experimental schooling scheme for girls, designed to counter the evils of female infanticide, ended in financial disaster in the Punjab in 1865.[38] In the north, conflict with the government over conversion and the teaching of a secular curriculum eventually resulted in the missions' retreat from any ambition to educate the village after just 17 years in the field.[39]

As in England, the "professional" Western bureaucrat was now marshaled by the state as he accepted a direct role in education funding and governance. In India, after 1871, Calcutta began using administrative fiat to "decentralise" its general administration throughout the Raj to save money. The aim was to maintain central control while allowing more Indians into junior administrative posts to justify higher local taxation levels. As a result, raising state revenue for education became more dependent on local funding via municipal and district councils, many of which were reluctant to disperse monies for education, preferring instead to spend funds on roads and local sanitation. The relatively uncontentious education cess mentioned above was replaced by five different and more visible categories of taxation that were designed to complement and cross-subsidize when necessary. However, famine and the determination to keep the Indian economy dependent and unindustrialized meant that taxation revenues by these methods were limited. Partly as a result of this, direct government education subsidies declined from 71 percent of total expenditure in 1870–1871 to just 26 percent in 1900–1902, and this also made the state dependent on the willingness of students to pay fees for their education.[40]

These changes to the administrative arrangements in the Raj also had an effect on the schooling curriculum. The limited transfer of rule to indigenous intermediaries, particularly in the lower law courts, meant that the state now directly encouraged the teaching of complex and culturally seamless civic duty values in the central schools. As part of this, European notions of citizenship were imposed on the school curriculum, especially in an attempt to fuse Hindu and Muslim law with the precepts of English Common Law in the lower law courts. Troublesome educators such as Gottlieb Leitner worked hard to create a vision of *Staatsidee* (state feeling); these moves later even proved to be a catalyst for the teaching of citizenship in schools at the metropolis and in other parts of the empire.[41] They also stimulated an opposing strategy adopted by disaffected, but increasingly organized, indigenous leaders. These leaders, especially those who were part of the Hindu polity, were now using British communications and print culture to coordinate their message, to be developed under the heading of "national resistance" in the twentieth century.

State Formation of a Reactive Indigenous Discourse, 1882–1886

To indigenous intellectuals, and to their countrymen generally, these stark state-sponsored interventions must have appeared naïve and illustrative of a barren Western education project by the late 1870s. Secret government translations of vernacular newspapers revealed a lively debate among them, particularly on the question of the medium of instruction in government schools. But most indigenous petitioners only found an effective voice at the extensive and impressive Hunter Commission hearings of 1882. This was because tightly controlled protocols of formal Raj reporting had previously excluded them from the official discourse.

The Hunter Commission was set up to examine why the aim of Wood's dispatch of 1854, to spread education to "the masses," had not been fulfilled. Hunter and his commissioners traveled to each province and presidency, where they held exhaustive public hearings and received lengthy petitions from interested parties. They also did an impressive job in ordering this information without too much official selectivity and censorship, to recommend important reforms. Foremost among these recommendations were those that applied to primary schooling. The commissioners concluded that state primary schooling should be taught in the local languages and the skills taught ought to be suited to the future life of most children and not just as the first step to the university. With this in mind, they expressed support for subjects that taught Indian methods of arithmetic and accounting. It was also recommended that state education embrace indigenous schools by training their teachers but interfering as little as possible with their curriculum so as to preserve their cultural values. Rather than relying on obdurate state departments to do this, the commission suggested that elected local bodies should be entrusted with the responsibility.

All this cost money. Hunter's recommendations for primary education alone would have required an increase in expenditure of 300–400 percent. Instead, provincial departments chose to divert the little extra money that was eventually granted by the state to expand secondary and collegiate education in the cities. This was even though only a small percentage of primary-school-age children were receiving an education by this time. However, some benefits did flow from the notice the Hunter Commission gave to primary schooling. Harsh modes of corporal punishment, especially in Hindu-dominated schools, began to disappear. The curriculum also broadened to include subjects such as hygiene, agriculture, and physical education. After 1884, state primary school class sizes grew whilst retaining their examinations. But smaller classes in the indigenous schools were preserved and children continued to have more discretion to learn at their own pace, even though more of their teachers were trained by the government.

Most significantly, the Hunter Commission gave voice to those not working under state auspices. Their arguments, touted mostly by an emerging nationalist intelligentsia, did not represent a paradigm shift compared to the earlier thought processes established or appropriated by the British. Rather, as the Hunter hearings showed, their views and schooling practices were now shaped by the dynamic already created by the state in India. For example, indigenous petitioners demonstrated a collective willingness to embrace important global imperatives including Western literacy, numeracy, and technical education of the kind already imperfectly imported by the Raj, even though they were stridently opposed to department-led education. These understandings later fed into national resistance narratives, and activist anti-British organizations such as the Arya Samaj demonstrated much greater success in recruiting students, even when imitating Western schooling models. Its schools and colleges embraced the teaching of girls and the teaching of English as a world language. As well, emergent *Pandha* and *Mahajani* schools offered commercial training to the sons of traders and shopkeepers. Even though some Muslim schools mimicked the government schools' primary curriculum, including the teaching of European-sponsored subjects, many educated Muslims viewed government schools as inferior to their own. Better-adapted indigenous schools also began to do more to provide the technical and basic literacy education of the kind that most agricultural parents wanted for their children.[42] And these moves partly anticipated Gandhi's Wardha scheme of needs-based learning in the early twentieth century.

ICS bureaucrats, who had so strongly influenced secretariat education decisions in each province in the 1860s and 1870s, had lost faith by the 1880s in ever being able to reach broad sections of the population. Their negative responses to the Hunter Commission clearly demonstrated this when the commission recommended that the government again attempt to reach village schooling using new, so-called *zamindari* schemes. With little extra money to spend, the India Office accepted ICS inaction in the 1890s in enforcing the implementation of the Hunter Commission recommendations, particularly those that endorsed primary schooling. It also rejected Hunter's private view that ICS control via departmental regulation should be reduced to allow for a more spontaneous connection with the general population and the kind of education it wanted.

The State Imports a New Generation of Patronage Appointees, 1885–1900

In the final phase of British-led education in India, a new generation of patronage-appointed educators began to work more independently of the government education department to bring about some educational reform. These educators had already received their professional training as teachers in England, Scotland, Ireland, and other parts of the "white" empire. The influence of these appointees was usually limited to the individual city-based schools in which they taught. But their actions were often also shaped by the educational thought of their academic patrons, who were responding, in turn, to modern schooling practices in Europe.

Two such patrons were Sir Joshua Fitch (Chief Inspector of Teacher Training Colleges in England) and Professor P. S. Laurie (Edinburgh University), who regularly won jobs for their students in India. Fitch generally recommended teachers whom he deemed were successful in the art of teaching in England instead of those with superior academic credentials. For example, in 1892, he recommended Herbert Knowlton as a "sort of School Board man" rather than four university-trained men to take up a senior teaching and administrative position in north India.[43] Laurie's patronage was also important in this period. His career was mostly concerned with raising the professional standards of teachers in Scotland by modernizing their training using new teaching techniques. One of his student recommendations for India was William Bell.[44] Bell was subsequently appointed principal of the Central Training College for Teachers in the Punjab, and his reforms reflected the ideas of his mentor. He established a new model school where student teachers observed each other team-teach in rotation for six hours a week. They were assessed using a new set of criteria that included assessment based on questioning skills, preparation, methodology and classroom pedagogy.[45] By 1912 this school was considered to represent best practice by British educators for teacher training colleges throughout the empire.[46] It was then replicated as the institutions for "training" teachers, run by the state, local boards, or the missions, grew rapidly to number 926 for men and 146 for women by 1921. This was in sharp contrast to the teacher-education scene in 1882, where only two government "training" colleges for male teachers of English were in existence for all of India and where "vernacular" teacher training was vested in a loose collection of neglected and underfunded normal schools scattered throughout each province.

Such patronage-driven appointments could not deliver men with the expertise necessary to embrace the indigenous school in the last two decades of the nineteenth century in the way that Hunter had envisaged. But they did bring a new generation of educators who were willing to test the departmental line in other ways, hoping the unique cultural background of the Raj still offered productive outcomes when educational innovation was attempted.

The End of British-Led State Schooling in India, 1901–1919

In 1901, the Viceroy, Lord Curzon, was the first senior agent of the state to identify the problem of an overly "decentralised" system that had permanently unhitched "the masses" from the Raj education agenda. He understood that the workings of the colonial bureaucracy were slowly dealing the British out of the education equation. Curzon's speeches as an interventionist conservative offended many Indian nationalists for their abrasive and patronizing style. But after he personally presided over a key education conference in 1901, mostly of departmental personnel, he introduced much-needed reform. This was too late to bring Hunter's recommendations, set down 15 years earlier, to fruition. However, Curzon's confident and energetic interest in education was enough to convince the India Office to substantially increase government expenditure on education. In 1901–1902, this funding stood at Rs 40,100,000 but, as a result of Curzon's initiatives, it was to increase fourfold by 1921–1922. This resulted in an increase in the number of primary schools by 59 percent and of secondary schools by 37 percent in the same period.

New arrangements were set in place for the universities as well. Expenditure on most universities up until 1900 had been only for a small office staff and for the administering of examinations in the colleges. Now that more funds were provided, at least some of these universities could take on teaching responsibilities. Curzon's Indian Universities Act of 1904 also tightened the rules for college affiliation and it provided for periodic inspection of all colleges. The act restricted the number of fellows of each university and decreed that 20 of them were to be directly elected. Furthermore, Curzon halted the policy of "withdrawal" whereby the state had earlier relinquished control over schools and colleges if private enterprise, usually the missions, had been willing to take its place. Instead, he introduced more controls on private schools and he argued that the government needed to run schools of all kinds in all parts of India as exemplars if overall standards of education were to be improved. Although he did not admit it, there was a second prong to this strategy, which was to monitor the growing political consciousness among the educated classes of India.

By the early twentieth century, contemporary authors such as Arthur Mayhew and H. R. Mehta still viewed British-led education as offering "progress" for the subcontinent with future expansion possibilities.[47] The growth in the number of schools encouraged this view. Despite Curzon's efforts, systemic schooling was already seriously weakened. Government and mission education in colleges and universities had largely become the preserve of the elites, as William Arnold had feared 50 years earlier. The disconnection between the state and the general population grew as government schooling was mostly in large urban centers and as the patrician Amateur Ideal ethic of the English Public school took hold in the wealthy colleges.[48]

When visiting India in 1912 Sidney and Beatrice Webb wrote disconsolately about the lack of interest by British officials in government schooling despite a growing demand for "popular education" among the general population. They saw British "administrative nihilism" and poor-quality government schooling as the main culprits for this apathy.[49] Others, such as Annie Besant and the theosophists, working without reference to the state, created a new ethic of Hindu/Western educational confluence for the oppressed. This was also at a time when the Indian National Congress was growing more strident and leaders such as Dadabhai Naoroji, C. R. Das, and G. K. Gokhale were demanding, among other things, that the medium of instruction in schools should be Indian languages, that history taught in schools should be of the subcontinent and not of the imperialist power, and that students be encouraged to think of their own Indian nation-state. The Calcutta University Commission of 1917, in a 12-volume report, identified further problems that no Britisher facing the politics of India in the early twentieth century could ever hope to address.[50] These included ongoing low levels of Muslim participation in state education and a recognition that only purdah schools for Hindu and Muslim girls could encourage them to continue their education into adolescence. The report also identified the problem of inducing students of higher castes to enrol in vocational courses, which remains a dilemma for India to this day.

Finally, the Morley-Minto reforms of 1909 were recognition on the part of the British that it was becoming increasingly difficult to justify vesting control over education, among other departments, in officials who were not elected by the populace. In 1910 the Calcutta government hopefully established one central education department for all of India. It also entertained G. K. Gokhale's ambitious plan for free compulsory primary education for all, but did nothing more. Only in the city schools of Bombay was the leading nationalist S. V. J. Patel able to persuade the Legislative Council to adopt this important principle.

Ultimately, it was to be the colonial state that was to retreat from the collaboration begun so effectively with its men and women on the spot, in the 1820s and 1830s, when it "Indianised" the Education Service in 1919. Indians resented the IES with its senior positions and much higher pay rates restricted to Europeans and paid for by the state from taxes raised in India. However, as inflation eroded these rates, it was the India Office that refused to entertain even higher compensatory pay rises. This greatly reduced the number of Europeans wishing to enter the service, even before Gandhi's first noncooperation campaign began in 1921.[51]

Instead, the colonial state moved from direct control of education to a "diarchy" model of governance (1921–1937), where education was one of the "transferred" departments responsible to the elected legislatures of each province and presidency. By 1919, large increases in population, as well as famine and inflation, had undermined the intent of many of the state's well-meaning education initiatives of the latter half of the nineteenth century. Underfunding, an undue emphasis on central schooling teaching in English, and the loss of many valuable indigenous schools had also resulted in only minimal increases in literacy. However, the workings of its educational administration, via the problematic agency of the provincial department, was a more important reason for the retreat by the British from systemic village-based schooling begun so hopefully in the mid-nineteenth century. In this sense, the failure of British-led education took place well before the Khalifat and Non Cooperation movements took credit for its demise. However, education under the Raj had produced many ethical and administrative questions that later generations would use to reference their own decisions on education. Understanding the dilemmas that state schooling posed for the subcontinent, experienced in the first instance by the British, also helped position Indian nationalists and their acolytes to take up the mantle of self-government 30 years later.

Notes

1. J. J. Rousseau, *Emile, ou, De l'education* (Amsterdam: Jean Neaulme, 1762); T. Day, *The History of Sandford and Merton* (Dublin, 1812).
2. C. A. Bayly, *Empire and Information* (Cambridge: Cambridge University Press, 1997).
3. F. Watson, *India* (London: Thames and Watson, 1993), 137.
4. L. Zastoupil and M. Moir, *The Great Education Debate* (Surrey: Curzon Press, 1999), 14.
5. East India Company Charter Act of 1813, section 43 (53 Geo. III, C. 155, s. 43).
6. Zastoupil and Moir, *The Great Education Debate,* 7.
7. W. Adam, "Reports on the State of Education in Bengal 1835 and 1838…" reprinted in *Reports on the State of Education in Bengal 1835 and 1838,* A. Basu (Calcutta: Government Printing, 1944), xxiii.
8. "Mr Adam's letter to Lord W. Bentinck on Vernacular Education," n. d., in [*W.*] *Adam's Report on Vernacular Education in Bengal and Behar Submitted to Government in 1835, 1836 and 1838, Rev. J. Long* (Calcutta: Government Printing, 1868), 47.
9. Adam, "Reports on the State of Education in Bengal," xxiii.
10. Basu, *Reports on the State of Education,* xix–xxiii. Adam's report was ignored when it was published three years after the Anglicist/Orientalist debates of 1835.
11. "Native Education Articles," Elphinstone Articles, MSSEur F.87/109.
12. J. Bentham, "Chrestomathia," in *The Collected Works of Jeremy Bentham,* vol. 8, ed. W. H. Burston and M. J. Smith (Oxford: Oxford University Press, 1983).
13. K. Prior, L. Brennan, and R. Haines, "Bad Language: The Role of the English, Persian and Other Esoteric Tongues in the Dismissal of Sir Edward Colebrook as Resident of Delhi in 1829," *Modern Asian Studies* 35, no. 1 (2000): 75–112.
14. Zastoupil and Moir, *The Great Education Debate,* 21.
15. "Minute Recorded in the General Department by Thomas Babington Macaulay, Law Member of the Governor-General's Council, dated 2 February, 1835," in *Selections from Educational Records 1781–1839,* pt 1, H. Sharp (Calcutta: Government Printing, India, 1920) 107–116.
16. "Resolution of the Governor-General of India in Council in the General Department, dated 7 March 1835," in *The Great Education Debate,* Zastoupil and Moir, 194–196.
17. J. P. Naik and S. Nurullah, *A Students' History of Education in India* (Delhi: Macmillan, 1974), 84.

18. "Minute dated the 2nd February 1835 by T. B. Macaulay," in Sharp, *Selections from Educational Records*, 109.

19. Lancelot Wilkinson was assistant resident in the central Indian state of Bhopal at the time and he wrote on developing a confluence of Brahmin knowledge with that of the West so as to "improve" India.

20. Zastoupil and Moir, *The Great Education Debate*, 60–61.

21. Brian Hodgson (1800–1894) was educated at the Fort William College, Calcutta. While holding several minor administrative posts in the East India Company, he was best known for his writings such as *Pre-Eminence of the Vernaculars; or the Anglicists Answered* (1837), which defended the orientalist cause.

22. Bayly, *Empire and Information*, 257–267. See also, L. Wilkinson, "On the Use of the Siddhantas in the Work of Native Education," *Journal of the Asiatic Society of Bengal*, 1834, art. 7, 504–519, cited in Bayly, *Empire and Information*, 257. See also, E. F. Irschick, *Dialogue and History. Constructing South India, 1795–1895* (Berkeley: University of California Press, 1994) and N. Dirks, "Colonial Histories and Native Informants," in *Orientalism and the Post-Colonial Predicament*, ed. C. A. Breckenridge and P. van der Veer (Philadephia: University of Pennsylvania Press, 1993).

23. Kazi Shahidullah, *Pathshalas into Schools* (Calcutta: Oxford University Press, 1987), 15, 25, 29, 33.

24. Rev. J. Long, "Brief View of the Past and Present State of Vernacular Education in Bengal," in Long, [W.] *Adam's Report*, 13–18.

25. For an analysis of Arnold's Punjab experiment see T. Allender, "William Arnold and Experimental Village Education in North India, 1855–1859: A Bureaucratic Precursor to the Development of State Education in England," *Revue d'histoire de l'education* (Canada) 16, no. 1 (March 2004): 63–83.

26. J. S. Hurt, *Elementary Schooling and the Working Classes* (London: Routledge & Kegan Paul, 1979), 10–15.

27. W. D. Arnold, "Memorandum as to a Central College at Lahore," January 21, 1856, no. 236, OIOC P/201/53.

28. M. Poovey, *Making a Social Body: British Cultural Formation 1830-1864* (Chicago: University of Chicago Press, 1995), 29–33, 42–52.

29. "Despatch from the Court of Directors of the East India Company to the Governor General of India in Council 19th July, 1854," no. 49, para. 14, in *Selections from Educational Records*, pt. 2, J. A. Richey (Calcutta: Government Printing, 1920), 368.

30. There was also provision for the establishment of three universities, one each at Calcutta, Bombay, and Madras.

31. T. Osborne, "Bureaucracy as a Vocation: Governmentality and Administration in Nineteenth Century Britain," *Journal of Historical Sociology* 7, no. 3 (September 1994): 60; P. Penner, *The Patronage Bureaucracy in North India* (Delhi: Chanakya Publications, 1986), 224.

32. Osborne, "Bureaucracy as a Vocation," 63.

33. Bayly, *Empire and Information*, 365.

34. These raw figures are drawn from J. P. Naik and S. Nurullah, *History of Education in India* (Delhi: Macmillan, 1974), 180–238.

35. R. Johnson, "Administrators in Education before 1870: Patronage, Social Position and Role," in *Studies in the Growth of Nineteenth-Century Government*, ed. G. Sutherland (London: Routledge & Kegan Paul, 1972), 110–114.

36. J. C. Bagal, *Radha Kanta Dev, 1784–1867* (Calcutta: Press & Literature Co., 1957), 47.

37. Census of the Punjab, 1891, p. 250, cited in K. Jones, *Arya Dharm Hindu Consciousness in Nineteenth Century Punjab* (Delhi: Manohar, 1989), 60.

38. T. Allender, "Robert Montgomery and the Koree Mar (Daughter Slayers): A Punjabi Education Imperative, 1855–1865," *South Asia* (New Series) 25, no. 1 (April 2002): 97–120.

39. T. Allender, "Anglican Evangelism in North India and the Punjabi Missionary Classroom," *History of Education* (United Kingdom) 32, no. 3 (May 2003): 273–288.

40. A. Misra, *The Financing of Indian Education* (London: Asia Publishing House, 1967), 186–191.

41. Leitner to Lytton, January 22, 1877, "Private," Lytton Collection, OIOC MSS.Eur. E.218, vol. 30, f. 23 (a).

42. "Evidence of Babu Hari Singh, Assistant Inspector of Schools, Lahore Circle," in *Report of the Provincial Committee for the Punjab of the Hunter Education Commission*, W. W. Hunter, 240; "Evidence of Sodhi Hukum Singh, Extra Assistant Commisioner and Mir Munshi, Punjab," in Hunter, *Report of the Provincial Committee for the Punjab*, 280; "Answers to the Questions of the Commission by the Anjuman Hamdardi Islamiya," in Hunter, *Report of the Provincial Committee for the Punjab*, 133; "Evidence of Lal Mulraj M.A." (offg. Extra Assistant Commissioner, Gujrat), in Hunter, *Report of the Provincial Committee for the Punjab*, 316.

43. Fitch to Maitland, November 28, 1891 [no procs. no.]; Fitch to Maitland (?) February 4, 1892 [no procs. no.]; Maitland to MacPherson, March 2, 1892 [no procs. no.] OIOC L/J&P/1828/91.

44. Professor P. S. Laurie (University of Edinburgh) to Sir W. Muir, March 25, 1885, no. 146, OIOC L/PJ/6/146. Moray House Records, Institute of Education, Herriot-Watt University, Scotland.

45. "Training Institutions for Masters Report 1886/7," Government College of Education Archives, Lahore, 62–63.

46. Lord Crewe to the Viceroy in Council, March 1, 1912, Crewe Papers I/6, no. 10, f. 2, Cambridge Main Library.

47. A. Mayhew, *The Education of India, a Study of British Education Policy in India, 1835–1920 & its Bearing on National Life and Problems in India Today* (London: Faber and Gwyer, 1926); H. R. Mehta, *A History of the Growth & Development of Western Education in the Punjab, 1846–1884* (Lahore: Punjab Government Record Office, 1929).

48. J. A. Mangan, "Images for Confident Control: Stereotypes in Imperial Discourse," in *The Imperial Curriculum: Racial Images and Education in British Colonial Experience*, ed. J. A. Mangan (New York: Harmondsworth, 1993), 11; J. A. Mangan, *The Games Ethic and Imperialism: Aspects of the Diffusion of an Ideal* (Middlesex: Harmondsworth, 1986), 124, 132.

49. "The Webbs in Asia," cited in C. A. Watt, "Education for National Efficiency: Constructive Nationalism in North India, 1909–1916," *Modern Asian Studies* 31, no. 2 (1997): 352.

50. M. Sadler, *Calcutta University Commission, 1917–19* (Calcutta: Government Printing Office, 1919).

51. C. Whitehead, *Colonial Educators: The British India and Colonial Educational Service, 1858–1983* (London: I. B. Tauris 2003), 18–20.

4

Disciplining Liberty: Early National Colombian School Struggles, 1820–1840

Meri L. Clark

During the revolutionary and independence period of the early nineteenth century, Colombians, like many other Latin Americans, committed themselves to creating a national education system despite the extraordinary obstacles they faced. In the 1820s, the new national government implemented its Enlightenment vision of universal education: schooling for at least two years for all Colombians, regardless of race, class, or gender. Many obstacles frustrated that plan: meager finances, deficient infrastructure, elite resistance, local hostility to centralization and secularization, and entrenched racism.

Education, specifically primary education, is a good site for investigating the question of the emerging nation's authority because the future of the republic rested, as it was seen in this period, on its citizenry's education. The new citizens of Colombia had to accept and participate in the project of education in order for the new nation to work, in the most enlightened of senses. If Colombians accepted that project, then they also supported the new government.

The Colombian state education project was not without controversy. This chapter explores several cases in which locals reacted to and against nationalized schooling and examines the ways in which those controversies developed. The local disagreements over state schools illuminate the larger controversy over the authority of the new nation-state. The first section discusses an alleged murder involving a teacher and his student. In this case, the local reaction suggests that the national government had gone too far by meddling with community politics from a distance. The central government had sent a teacher who had not been vetted by the community and had granted that teacher too great a measure of authority over the children and, indeed, the adults of that town. Yet in the other cases examined here, black and indigenous Colombians called the government on the carpet for failing to do enough to develop school infrastructure and promote egalitarianism within it. These locals claimed that the Colombian government did not actually support the racial and social equality that its own "enlightened" rhetoric vaunted.

The local impulses to criticize the government appear contradictory, but they link together. Colombians of the postcolonial era both feared the resurgence of despotic authority and worried deeply about social disorder in the absence of a strong central government. These fears divided Colombians on governance issues in this early national period. Early national education reflected this tension about state authority: How much disciplinary power should a teacher enjoy in the classroom and outside of it? Who decided which children in a community would attend school and with what funds? When was race or ethnicity the determining factor in a decision to enroll, educate, or graduate a student—or not to—and when did it involve "merit" or "virtue" alone?

Early national education showed the ways in which Colombians muddled through the difficulties of establishing a new, independent government and building a republic, person by person. After all, if the ties to the Spanish Empire had been rent asunder in the 1810s, so too were the clothes and social graces of the colonial period. What, then, was the comportment of a Colombian citizen? What clothes did this man or woman wear? What style of speech did he or she employ? How would these citizens be taught to think, speak, behave, and dress? Here, too, lay a source of conflict. Some Colombians preferred to maintain the older ways of the empire: the same teachers; textbooks; and rules of access, dress, and behavior. Others wanted to uproot the old system of hierarchy, with its racism and elitism, and replace it with an egalitarian republic of citizens. Neither extreme stance held for long; certain political compromises materialized as local concerns intersected at the national level. But when the intersections became clashes, opinions again hardened about the nature of virtue, the merit of racial equality, and the fate of the republic. Education, especially of the youngest Colombians, was a site of conflict as intense as family emotions, financial worries, and political opinions, since these often intersected. In the early national period, Colombia teetered on the brink of renewed civil war. Early national education debates illustrate how Colombians fought each other over local matters, just as much as they fought to remain united under the aegis of the republic.

Colony to Nation

The wars of independence had laid waste to vast regions of Colombia. Colombian schools—already very few in the late colonial period—were now shuttered, reflecting the wartime devastation. The nation-building efforts of the 1820s in part focused on reopening, improving, and developing the country's schools. Colombians of very different political stripes linked their notions of freedom and democracy to universal education. The first independent government, led by President Simón Bolívar and Vice President Francisco de Paula Santander, promoted the idea of universal elementary schooling. By national law in 1821, all Colombians, regardless of age, gender, or race, were expected to attend at least two years of primary school to achieve basic literacy and knowledge of their civic duties. Developing the first national school system in Colombia was an erratic and contested endeavor.

Until independence, most colonial schools were housed within, and operated by, various branches of the Catholic Church. Without an established parish church or mission school, some colonial town councils (*cabildos*) struggled to run a small school with a short-term teacher. If other public expenses such as road repairs or postal services had exhausted a town's treasury, then wealthier residents or the landed elite of an area might volunteer to subsidize the school rent and the teacher's salary. Unsurprisingly, the independence wars had decimated large estates (*haciendas*). Many landowners left the country; their absence deprived poorer neighbors of accustomed financial aid. Financial difficulties persisted, or worsened, in the early national period. Many towns could not afford a school, so they simply did not open one. In such cases, frequent through the 1820s, the national government's sole, feeble response was to demand an explanation for the town council's delay and order it to open a school immediately. The councils answered unanimously. Hundreds wrote to the secretary of the interior, José Manuel Restrepo, in charge of the nationwide school system; each council described the ravages of war on their town, the destitution of their people, and the dearth of funds in their treasury.

Teachers had scant resources to operate schools if they did exist. For example, after 1821, the state expected teachers to use a standard primer to teach spelling and writing. It ordered thousands of primers to be printed and distributed to every cabildo. However, posting the materials involved great delays and expense.[1] Purchasing books and school supplies also strained budgets. Paper, pens, and ink were scarce and costly. Writing quills had to be carved by skilled hands, a task that fell to the teacher or an advanced student. Many teachers avoided the chore because they had no quills to sharpen. Then students wrote on chalkboards or, following Joseph Lancaster's innovation, they fingered their lessons into slats of sand. Although sand was the most rudimentary of Lancaster's requirements, many villages still could not supply it.

Sending teachers to remote and extremely impoverished villages became one of the young republic's most daunting challenges. Rural areas had always faced obstacles to education.[2] Even when villages could afford to employ state-trained teachers, they still suffered lengthy delays waiting for him and his supplies to arrive from larger cities. Itinerant or "circuit" teachers were more common. These men traveled a circuit of several villages, working in each place for one to three months. Circuit teachers had unstable work because their constituencies usually could not pay or even house them. In lieu of wages, villagers might provide a teacher with a sleeping mat and a subsistence diet of plantains, yucca, eggs, or dried meat.

Rampant poverty coupled with a dearth of teachers even in urban areas in the early national period. Many colonial teachers had been priests, nuns, or laymen working under the aegis of the Catholic Church. Ironically, state school efforts in the 1820s dispossessed many teachers of their classrooms and further disrupted school operations. Partially in response to cabildo complaints, the government took complementary measures to support state-run schools. One law secularized every minor convent (*convento menor*) staffed by fewer than eight priests or nuns. After this controversial law dispossessed many clergy, the government tried to reopen the smaller convents as schools, poorhouses, and hospitals. A few of the largest convents continued to teach adults and children biblical history and Catholic doctrine. Adults and children in larger parishes thus could still attend mass and religious history classes on Sundays, while children could meet up to four times a week with clerical teachers. But the early national state withdrew any other support for conventual education or proselytizing efforts such as mission schools.[3] The state intended to dismantle Catholic Church power in Colombia, but it simultaneously undermined its own efforts to create a national school system. Without the clergy, who remained to teach Colombian children?

Many colonial teachers had been fired because of new licensing procedures. The government had intended the licenses, in part, to shield teachers from criticism of their competence and morality. But licensing caused more grief at the local level. Poor towns were less likely to recruit teachers, so the government's demand that cabildos guarantee a teacher's reputation and education restricted recruitment to an impossible point. In some instances, long-working teachers were dismissed because they did not meet the new licensing requirements. Other colonial stragglers did not suit the changed political climate of the profession. For example, *maestro* (teacher) Agustín Torres had taught in Bogotá since the early 1780s. Even after the Spanish Empire imploded in 1808, Torres continued to teach. At different times during the revolutionary period, royalist and revolutionary soldiers used his school as a garrison. When Torres considered these sacrifices, he said, he could only protest being driven out of his teaching position after Vice President Santander began his educational reforms in the mid-1820s. After several failed attempts to be reinstated as a teacher, Torres finally gave up and fought for a pension instead. Despite his advanced age and long service, the government would not forgive Torres's work as a teacher under the Spanish Empire. Many teachers of the colonial era were similarly dispossessed of their students and livelihoods; unlike Torres, they engineered reappointments, although usually to remote, rural schools.[4]

The overturn rate for teachers across the country was extremely high. Teachers quit or were fired in such rapid succession that their employment can be tracked only with difficulty, even when records were kept by the national or local governments.[5] When the hiring difficulties abated for town councils, fresh complaints about teachers and their schools flooded the council offices.

State-appointed teachers bore the brunt of local complaints, which made teacher recruitment even more difficult. Colombians resented what they perceived as the moral breakdown of their society spearheaded by the state's secularization efforts. Questions were raised about the prudence of many laws in the early national period. The increasing tax burden led many, especially the elite, to critique the recently introduced Lancasterian system. Some elites doubted the pragmatism of educating the poor along with the rich. Many others worried about the cost of building and maintaining Lancasterian schools, which appeared to provoke the resentment of the poor majority regardless. Most Colombians did not direct their concerns with these centralizing laws directly to the national government. Instead, they launched attacks on local personalities and institutions. In particular, Colombians criticized state-appointed teachers, many of whose names and reputations were unknown because they had been assigned to distant schools by the national government.

Financial and moral anxiety triggered the challenges to the authority and morality of state schoolteachers. Townspeople worried about the teachers' morality, training, and honor, as well as the teaching methods they employed. Doubts remained especially about the mores of state-licensed teachers and whether they could be entrusted with the care of children. How could townspeople judge a teacher's character if he were sent to them by an unknown officer in the distant provincial government? The central government, and its emissaries, occasionally provoked Colombians, who guarded jealously their autonomy over local affairs. But state interference was only one part of the problem for some nettled Colombians. Teachers' authority, their character, and their personalities, came under fire as well. The following case examines a case of extreme hostility toward a public school teacher in the early national period and analyzes the reasons for his condemnation and eventual exculpation.

Local Authority in Primary Schools

Few archival documents detail particular concerns that townspeople had with a teacher's character or conduct, although many testify to their financial difficulties. In one rare case, a town revealed its struggles to understand a teacher's real authority and to curb that of its own government officials. In 1826, at least one official, and possibly several of the residents, of the town of Honda registered a serious charge against Bartolomé Guerra, the teacher in their recently founded Lancasterian primary school. The complaint expressed general concerns with state-mandated teaching practices and the teacher's role in the community. More gravely, maestro Guerra had been accused of committing "excesses in the exercise of his duties."[6]

Guerra had murdered one of his students, stated the allegation. "Worthy individuals" claimed that he had punished one of his students, placing the boy face down in stocks and causing fatal injuries. News of the child's death in the stocks shocked the people of Honda, but they were most distressed by the rumor that the teacher had regularly applied "this horrific punishment to all of his students." The complaint also stated that Guerra had "forced others to stand with their arms spread in the form of a cross, holding heavy stones in each hand for the same amount of time, all because they hadn't complied with their obligations to learn or because they had been a little tardy to school or because they were distracted in class when they were supposed to be studying."[7] The official suggested to his readers, the town council, that Guerra had also blasphemed. The punished students' arms took the shape of the cross, which implied that Guerra had figuratively crucified them. In this light, Guerra was not just a murderer but unchristian too.

The cruelty, duration, and frequency of Guerra's punitive measures provoked anger and disgust. The town official demanded that the Colombian secretary of the interior "dictate an active and energetic measure that would contain the barbarity of this preceptor [who is] so alien to good philosophy." If not, he threatened, all the parents would soon remove their children from Guerra's influence. Then the teacher "doubtlessly would have ruined the Youth and the school, and made the children of Honda abhor this kind of education, up to and including the very word *school*."[8]

One month after the initial allegation, a more senior town official (*el Jefe*) said the charges against the teacher were "absolutely contrary and foreign to the principles and human sentiments that distinguish this Individual."[9] The Jefe found no evidence for the murder allegation: Guerra's school did not have tools of corporal punishment, and no one had ever seen him punish students with stocks or stones. The town treasurer, town councilmen, and several *vecinos* (tax-paying residents), including the Jefe himself, inspected the school regularly. During the year or so that Guerra had taught in Honda, he had only ever punished children by putting them in "a small room of correction" for a few hours. The existence of this detention room was not Guerra's responsibility either, since it had been built into the school. Aside from that, the only "light punishment he does impose is to make them kneel; their arms are not opened in the form of a cross, nor is the whip, the Ruler, or any other punishment of terrorism applied."[10] Since the teacher employed moderate punishment, the Jefe argued, he was neither excessively cruel nor wholly permissive. The battle between these town officials involved the measures to be included within the definition of "moderate discipline." The deeper debate involved who held the authority in a town to determine justice and the limits of punishment.

Rumor played a large role in shaping this controversy over Guerra's character and the limits of his authority. Public opinion also influenced the deliberations of town officials. For example, the Jefe had heard from several townspeople that the dead student, Felix, had just returned to school after three days of unexcused absence. This irresponsibility prompted maestro Guerra to punish the boy. It was said that Guerra had made him stand in the shape of a cross before the class, but the boy began to jump around, which irritated Guerra even more. He told Felix that "if he wanted to be a Clown, he would have to stand on his head." Rumor had it that Guerra made the boy do just that, supported by the wall, for several minutes.[11] The punishment complete, Guerra then asked Felix to bring an excuse for his absence from his mother, who, coincidentally, had three other children enrolled in the school. The mother sent a note explaining that Felix had been ill with a severe headache and fainting spells, also commenting that another of her sons and her mother had experienced the same symptoms and died recently. "From this point, Mr. Governor," the Jefe recorded, "the ignorant populace confused the two brothers [and] assumed that the dead child named Gervario was in fact Felix, the one who had been stood on his head at School."[12]

Rumors ran rampant, confusion prevailed, and the townspeople fretted. The town's doctor, Dr. Gonzalez, added his own perspective on the rumors and the reality of the situation. He thought that several factors exacerbated the generalized confusion and anxiety.[13] First, the official who made the allegation had been overly zealous and "unreflective in his haste" to accuse the teacher of wrongdoing. Second, Dr. Gonzalez had been called to examine a boy (Felix's brother, Gervario) when a severe fever had not abated; the doctor saw no hope for the boy's recovery and simply tried to comfort the mother.[14] The doctor speculated that Felix's family was one of the first carriers of the epidemic disease that had afflicted the town in recent weeks and caused much anxiety among the residents. The mother was the third factor in the general confusion: not only had she just lost her mother and her son, she was afraid that another son would soon die. She told Dr. Gonzalez that she worried that one of her surviving sons, Felix, would succumb to the fever since he had been punished in school the day before, but the doctor reassured her that Felix was healthy. The fourth factor soon materialized. Maestro Guerra himself asked the doctor to examine Felix because a rumor had spread "among the common people that the Young Man who had been stood on his head had died as a result."[15] Although the teacher was convinced the rumor was false, he did not want "his Conduct to be stained and, no matter what the case, he wanted to receive an authentic Document from a physician about the matter." The doctor, the prosecutor general, and several important residents then accompanied Guerra to the student's house, where they found one son (Gervario) dead and the other (Felix) in good health. With that, the senior town official closed his investigation.[16]

So why did the "ignorant populace" react so quickly and forcefully to the news of the student's supposed death? Clearly, the notion of a child dying at the hands of a teacher would have alarmed the town. The commoners in town, unlike the elite, might not have spent enough time with Guerra to gauge his character. Or he might have provoked suspicions in them earlier for other reasons.

Suspicions ran as high as the fever in town. Yet most of the struggle over the teacher's actions and character took place on paper, between town officials and the local elites. The first official to make the murder allegation against Guerra was himself accused of laboring under "a heated imagination." The accusing official had insulted the honor of the teacher *and* the town authorities, the other elites said. Dr. Gonzalez wrote that every official had taken a "sacred oath" to act as a "zealous defender of minors, a spokesman for the Public and, above all, a supporter of the Laws and public defense." In this case, a child was not threatened, and an official had committed an egregious error by presenting unsubstantiated charges to the national government. The doctor reminded the readers of his report—the governor, secretary of the interior, and vice president—that there were honorable men in Honda. These officials "would not have let pass unpunished an incident as atrocious as that which [the accusing official] supposes happened, even if the Attorney General and the Father General of Minors (*Padre General de Menores*) were to have demonstrated a criminal silence in the matter."[17] A few weeks later, the provincial and national governments upheld the ruling that the teacher was innocent.[18]

The murder of a student was a shocking allegation. This explains, in part, why maestro Guerra's case advanced to the upper echelons of government and why it sparked such anxiety among the people of Honda. But what about the teacher so outraged the minor bureaucrat who initiated the charges? A personality conflict might have spurred the bureaucrat's allegations against Guerra, or at least intensified the emotions he and the nonelite townspeople felt about the murder inquiry. Townspeople quickly spread the rumor that the teacher had abused his power in the classroom. Even if it were a lone official who first stoked the fires under Guerra, he found ready kindling among the community. Maestro Guerra did not seem to enjoy the trust of "the common people," given the sudden and vicious scuttlebutt. He received his appointment to teach in Honda after training in the country's distant capital.[19] As an outsider, Guerra might have been a more dubious authority figure for the townspeople. However, the murder charges brought against Guerra also figured in a broader struggle over teacher authority in national schools.

Several points of tension soon developed between national and local opinion about the school system and teacher autonomy. Another joint under extreme strain involved the teacher's right to discipline students. More broadly indicted was the martial character of the Lancasterian curriculum in public schools. In 1820 and 1821, the national government proscribed school punishments such as whipping and splinting students, except for "the rare occasion when the child's defects denoted depravation."[20] The state told its teachers to administer punishments according to the "temperament and inclination" of the child. Under the Lancasterian system then operating in Colombian schools, this precept meant that teachers could make children stand silently for hours in front of the class or recite moral axioms until school recessed. However, teachers could not use the "tools of terrorism" that Guerra was accused of employing to discipline students. The state empowered teachers to chasten students and allowed them to scrutinize the behavior of parents as well. Just as the state expected students to behave well in class, it expected parents to govern children judiciously at home. Thus, if parents refused to enroll their children in school, they could be fined and, in the most extreme cases, jailed. The state expected teachers to form judgments about parental recalcitrance or willingness to aid in the nation-building process by educating their child. In this sense, teachers could shape the fate of the adults in their community along with the children's.

Teachers' power within the classroom became more important as increasing numbers of Colombian children attended school. Although lashings had been banned, no law limited teachers' authority to punish students.[21] The lacuna left room for interpretation of, and conflict over, teacher authority inside and outside the classroom. Legal attitudes about the whip, at least, had changed in the courts of Latin American republics.[22] Teachers who practiced "new" Lancasterian methods tried to find common ground with parents who were accustomed, if at all, to the "traditional" methods of their parish priests. For example, teachers grew more inclined to punish students with seclusion, since it involved physical confinement but no contact. However, many continued to make students kneel and receive blows on their hands with the *palmeta* (rod).[23] Censure and support for corporal punishment fluctuated over the years, and varied across the country. Although the cabildos of Bogotá, Caracas, and Cartagena had banned the most severe forms of corporal punishment by the early 1820s, there is no evidence that these decrees were duplicated in the surrounding rural areas, or even that corporal punishment actually ceased in city schools.

Ethnicity and Economy

Poverty limited educational access in Colombia, and it carried moral and racial connotations. Indigenous and black Colombians faced more obstacles to entering schools than did their white compatriots of the same economic situation. Vice President Santander said that it was of the utmost importance to "remove the indigenous people of Colombia from their state of misery and ignorance, which derives from the Spanish legislative system," and which could be improved by the powerful medium of education.[24] To counteract some of these obstacles, in March 1822, Vice President Santander ordered every *colegio seminario* (secondary school) in the country to admit indigenous students and he created a scholarship fund for four qualified indigenous boys.[25]

In principle, the Colombian government supported Indian education. In reality, it undercut the economic and political power of many indigenous communities to support state schools, usually intended for white elite children. The dilemma for the Colombian government was this: postwar projects such as road and bridge construction often took precedence over schools. The government allowed poor Colombians to exchange their labor for tax payments—building a bridge rather than trying to procure cash for taxes. But so many Colombians offered their labor to avoid paying taxes that the government needed other tax revenue to support schools. So the government turned to indigenous communities and used revenue from household taxes and the sale of reservation lands. For example, the Moniquirá reservation lands were sold to fund a primary school in the town of Villa de Leyva in 1829.[26] The government claimed that the reservation and its people were destroyed (*un pueblo destruido*): no more than 80 indigenous households remained, the majority headed by women. From their larger parcels of land, the Moniquirá people rented "arbitrarily the remainders . . . for small amounts." The government thought it could manage the lands better, and planned to seize the Moniquirá lands to rent them at a higher rate.[27] The government was unconcerned that the 80 families would be displaced from their lands. One official wrote that he hoped "those same indigenous people could find use in attending the school in Villa de Leyba, unless their plots of land lie a great distance away."

State seizure of indigenous communal lands has been considered part of the "liberal attack" on indigenous rights in the early national period. But this assault must be understood in the context of the national promotion of indigenous rights, rhetorically and realistically, in the Colombian republic. Universal education and indigenous rights were significant elements of the platform of republican ideals, even if reality so often disappointed.

The indigenous and white people who lived near and worked in the salt mines (*salinas*) in the central highlands of Colombia provide an interesting perspective on government centralization and local autonomy.[28] The relative stability of salt production and revenues made the towns of Zipaquirá and Nemocón important, but politically troubled, economic centers.[29] The highland peoples voiced their concerns to town governments through civic protest and uprisings; but they also used the established political channels of the government. From independence onward, their attention focused on the points at which local and national interests diverged—notably, primary schools.

Zipaquirá's primary school first entered the ambit of national government during the brief existence of New Granada's earliest independent republic (1812–1814). In 1813, Mayor Manuel Bernardo Álvarez and the friar who taught at the school, Felipe Guirán, asked the executive government for a new schoolhouse.[30] The mayor complained that the old thatch barn currently serving as their primary school was uncomfortable for the teacher and the students and it lay far from the center of town. To make matters worse, the barn's owners had been demanding its return for their own use.[31] Alvarez argued that education was important enough a matter that the teacher be allowed to use the office that once served the *aguardiente* (cane liquor) tax collectors. Even if this move was temporary, a better room would help the town "serve its pressing duty" to educate its children. Alvarez himself guaranteed to keep the space clean and free from damage. Friar Guirán wrote separately that he had offered litanies of praise and honor to the saints and that he trusted in God to aid "this Holy Community in its present need."[32] In early January, President Primo Groot agreed to let them use the tax office for school, which was used through the revolutionary period.[33] Zipaquirá's substantial salt revenues became a vital issue for local and national government officials.[34] Regional divisions had strained the frail economy and tenuous national governance. Zipaquirá's conflicts revealed the strength of local political will.

Colombian elites opposed public schools for financial and religious reasons. For example, in 1823, the famed Lancasterian teacher José María Triana took up his new post as maestro in the salt town of Zipaquirá. Soon afterward, he protested his pitiable pay.[35] The governor had asked vecinos[36] to contribute 200 pesos each year to the teacher's salary, to make it the "reasonable" sum of 400 pesos annually. But the vecinos opposed any municipal tax that did not fund the parish church. Ever since church construction began, the vecinos claimed that their resources were scarce. If the government asked them to contribute another 200 pesos a year, they would bankrupt themselves. The mayor of Zipaquirá had long failed to persuade the townspeople to pay taxes for projects they did not support. He recommended abandoning the school project.[37]

The Zipaquirá town council tried to find a middle path. Faced with such fierce elite opposition to the school and, at the same time, with the national government's demand that it remain open, the town council suggested diverting revenues from the salinas to pay the teacher.[38] If the state imposed a separate tax for the school, the vecinos would consider it "repugnant" and refuse to pay. The Zipaquirá elite denied the public teacher his salary to protest the national government's trespass into local matters. They disagreed deeply with the broadened scope of the centralized state authority.[39]

In this precarious political climate, the national government refused to nego-tiate with a cabal of powerful citizens. It feared the loss of legitimacy if it were to acquiesce to local demands. From the perspective of the Colombian govern-ment, Zipaquirá had flouted the authority of constitutional law. "The People of Zipaquirá, even less the *Vecinos* singularly," declared the government, "do not have the authority to revolt, resist, or protest against the laws."[40] The secretary of the interior, in charge of schools, ordered the town council to pay the teacher without recourse to the salt revenues.

In contrast to Zipaquirá, indigenous people played a much more significant role in battling elites in the other highland salt town of Nemocón.[41] The Nemocón Indians battled the town's new teacher to start a school for their own children. In March 1823, a young man named Luis Vargas de Tejada started as Nemocón's only teacher. A few months later, he reported many frustrations with his work. Vargas's distress, evident in his letter to the interior secretary, offers a rare glimpse into the personal problems early national teachers faced in their work and life. Vargas had come to Nemocón as a young man of 20, eager to work on behalf of the "wise and philanthropic views of the Magistrates that pro-moted the development and establishment of mutual teaching schools."[42] Immediately, three obstacles thwarted his desires: the parish had no house in which to teach, the vecinos did not pay half of his salary (400 pesos), nor did they pay for the supplies required by the Lancasterian method. The manager of the Nemocón salt mines, Rafael Morales, was the only person to offer aid. Morales "happily relinquished . . . the only habitable residence in this Village," a two-room building on salinas property, in which the teacher could live and work temporarily. In two separate town meetings, the mayor urged the towns-people to contribute to the school, but most claimed to be too poor. "They envisioned only the hardship of the expenditure," the teacher lamented, and "not the utilities that their children's education could bring."[43] Given the town's resis-tance, the mayor decided to pay Vargas from the revenues of "the ancestral lands of the Indigenous community," some of which served as Nemocón's grazing land.

Just when Vargas thought his salary dilemma had been resolved, his school was beset by new troubles. First, a pair of British businessmen (*empresarios*) named Johnson and Thompson leased the salt mines. So Vargas expected himself and his family to be evicted from the schoolhouse, since it belonged to the sali-nas. But the teacher thought the village houses were "entirely too small, damp, and unhealthy" for his mother and his sisters to inhabit. Before he abandoned Nemocón entirely, Vargas appealed to the "noble sentiments" of the provincial authority and asked for help. He hoped that the government would ensure that Nemocón's children "would not be deprived of the most valuable fruit of our Liberty when they have just begun to receive the lessons that should dispose them to true and solid learning." Vargas asked the provincial government to negotiate with the new British managers on his behalf, to let him and his family live and work in the salinas house as Rafael Morales had done.

Maestro Vargas was unaware of—or perhaps indifferent to—the second pro-blem that affected his plans—the Indians of Nemocón themselves. The indige-nous group demanded accountability and action by the city government on their behalf and on behalf of their children. Rafael Morales, the salinas manager, informed the provincial government that the Nemocón Indians had come to him to claim their rightful share of the salt revenues.[44] The indigenous community knew perfectly well that 4 percent of the salinas revenue had been destined by law to serve their needs. They knew, too, that in 1823 their portion of the salt coffers should have amounted to 80,000 pesos. The problem was that some of the Indians' money had been "lost by private persons" and that the First Republic of Colombia had spent another portion during "the time of the political transformations" in the 1810s. Only 40,000 pesos remained from the Nemocón Indians' share. None of it had ever "provided a single benefit to the Indians of this Village." A few years earlier, the Nemocón Indians already had petitioned the town council to build them a school. They also asked the council to apply revenue from their communal lands (*resguardo*) to establish grants for three Indian children to attend school in the provincial capital. But the town council "neglected the proposal because of the disturbances then." Having been ignored for years, in 1823, representatives of the Nemocón Indians reiterated their complaints. They protested that the new teacher had used their revenues to fund his school while they still had no funds of their own.

This time, the Nemocón Indians had an advocate. Manager Morales claimed that times had changed. He thought the indigenous school was viable. The national government had promised to "foment the culture of the Villages and the education of the Youth" and guarantee "the rights of the Indians of this parish to their property."[45] So Morales promised to allocate a portion of the Nemocón Indians' salt revenue to endow a school for the indigenous *and* white children of Nemocón. The schools would not be subject to "the changes that are in store for the salt mines," nor would they provoke "the contingency and antipathy" that stemmed from individual taxation.[46]

But Indian education was pushed to the wayside once again. The secretary of the interior flatly refused to negotiate with Manager Morales about the Nemocón schools. He ordered the town council to endow the parish school, but not the indigenous school.

The secretary of the interior was the main educational policy maker at the national level. In the first decade of Colombian independence, the secretary maintained two general stances in regard to education: he wanted towns to finance their own schools and to hire teachers trained in provincial capitals. The secretary of the interior expected town councils to support teachers who had been assigned to their schools, even if the townspeople rejected the person or the very idea of the school. Tensions arose when wealthy, white vecinos refused taxation to support schools for the poor and nonwhite children of their districts. Schools offered a short-term child-care facility for working families with no kin or neighborhood networks to assist them. Urban and rural families usually depended on children's labor, so absenteeism was a problem in many schools. Despite their problems, public schools became more popular among nonelite and elite Colombians as in the early national period. School enrollment—a tiny percentage of the country's entire population in the 1830s—grew at a creeping pace, but it did grow.

Race and Ethnicity in Schools

Even though a greater number of and poorer Colombians aspired to send their children at least to primary school, the "enlightened" legislation that had promoted universal education often failed to advance their goals in practice. The poor encountered institutional indifference when trying to enroll in school, while black and indigenous people faced more institutional and social resistance to their ambitions. Poverty simply magnified the racial and ethnic barriers to access. Although no schoolteacher, superintendent, or government official recorded the perceived racial categories of Colombian students, my study of the Public Instruction archives suggests clearly that whites and mestizos dominated the ranks of primary and secondary students. However, it is also clear that, in regions with large populations of freed slaves, such as the Caribbean and Atlantic coasts, black Colombians had improved chances for social promotion through military service, education, and employment. In the highlands of the country, indigenous communities sometimes held the political and economic clout to protest government action, or inaction. Even if the successes seemed few and the achievements minor, Indians and black Colombians did voice their demands for education in the early national period. Their claims became the groundwork for late nineteenth-century developments toward egalitarian schooling.

In the early national period, the government maintained that education was the path toward national unity and stability. This rhetoric suggested that marginalized indigenous and black peoples would be incorporated into the national imaginary. The everyday reality of national incorporation meant the elision of cultural differences through policies of assimilation.

In schools, indigenous and poor children confronted policies that limited their cultural expression. In 1837, for example, the Pamplona school board introduced regulations designed to "better monitor matriculated students and their courses of study."[47] The primary school students learned the "rudiments of reading and writing in the Spanish language and the Christian doctrine," but supplies were scarce—the school lacked benches, tables, and chalkboards; many catechism readers had disappeared. Of the 80 boys enrolled in the school, only 50 or 60 attended regularly. The school board concluded that attendance was so low because the "poor parents did not want their boys to attend so they could avoid having to keep them tidy and clean,"[48] that is, the board thought that the parents were too poor to clean and clothe their children for school. More likely, the absentee children were at work.

The girls' school fared better in terms of hygiene, but attendance was poor there too. Fifty or sixty girls regularly attended to learn the rudiments of writing, sewing, and Christian doctrine. But their principal had taught "only two" girls how to embroider. The school board intended to remind the principal of her "obligation to instruct them in the labors natural to women, being the principal branch of education of the fair sex (*bello sexo*) and in compliance with the law of studies."[49] The lack of embroidery lessons overshadowed the fact that the girls still did not have copies of the state-mandated catechism and primers for reading exercises.

The Pamplona school board also criticized students wearing *ruanas* (heavy woolen capes). The board had been informed that, "despite the measures taken to make the young people attend scholastic functions with due decency," students had continued to wear ruanas, attire which was "contrary to tidiness and cleanliness."[50] Colombians of various ethnicities and social classes appreciated ruanas for their durability and resistance to the chill wind and rain of the Andes. But for the elites, the ruana indicated that its wearer regularly labored outdoors or traveled on foot. It marked a person as a member of the uneducated indigenous, black, or racially mixed lower classes. Since the ruana connoted a lower racial and class status, the school board considered it inappropriate for students to wear in school. The board encouraged teachers to punish students for the "abuse" of wearing ruanas to school.

In the eyes of the white elite, the nonelite racial majorities were destitute in many ways—financially, culturally, and morally. The Pamplona school board revealed a general elite disdain for lower social orders. More importantly, its analysis of absenteeism and ruanas showed that it linked together hygiene, willpower, and moral decency. In the elite formula, cleanliness equaled decency, decency marked a strong will, and strength of will meant strength of mind. Thus, unkempt or threadbare clothes betrayed a child's—and the parents'—weak moral and intellectual resolve. This early connection between culture (including dress and behavior) and morality figured significantly in developing educational policy in the positivist era.

Behavior—or "conduct"—was an important marker of social status in Colombia. Though easy enough to observe, classifying indecent behavior often resulted in quarrels, even lawsuits. Conduct mattered because social status was also laden with racial connotations in Colombia. Just as a ruana might signify to a white elite Colombian that its wearer was a poor Indian, a vulgar word or rude gesture might mean that a person was uneducated and racially mixed. Similarly, in school, teachers made direct connections between students' conduct and manner of dress and their moral character and racial category.

Morality was commonly attributed to physically discernible traits in early nineteenth-century Colombia. Assessing a person's character was fraught with perceptions of race and social status that were specific to a town or an entire region. In 1826, for instance, Joseph Álvarez charged the academy of Caracas with discrimination against him and his son, a student there, who were both *pardos*—black or racially mixed people primarily of African descent. Álvarez, a landed resident of the small canton of San Carlos, claimed that his son had been forced to leave the Caracas Academy because other students and their parents had racially harassed him. He petitioned the Venezuelan intendant J. Manuel Landa to reinstate his son.

Rebutting the charges, Rector Joseph Nicolás Díaz wrote that the academy had always complied fully with the law, which had abolished "the distinction of classes that once existed; the Academy had opened its doors to everyone . . . and had omitted any question about a potential student's lineage."[51] Álvarez's son had attended the school for two years, which Rector Díaz thought sufficient time to graduate from the institution. Twenty-one-year-old Álvarez had attended schools in his hometown of San Carlos and then in Valencia, before finally enrolling in the Caracas Academy. Díaz wondered why, after these years of study, Álvarez had still not managed to test out of the primary school levels. Moreover, Díaz alleged, once Álvarez had entered the school, he began committing offenses against the entire school community. Díaz did not specify the nature of these transgressions, but he attributed Álvarez's behavior to his years of apparently fruitless study. The rector had taken measures to correct Álvarez's mistakes, but it seemed to him that the student's negligence of his studies, his "harmful conduct," and his "incorrigibility" merited his expulsion.[52] His decision, the rector averred, was not based on race. As Díaz wrote to Intendant Landa,

> No *pardo* has ever entered the Caracas Academy without wanting to subject himself to the labors that learning requires. You sir, know well how many of them live in this City. Many of them attend the school's Lectures and there is even one from the very same town of San Carlos who holds a grant, which he is generally considered to deserve, given his worthy and normal comportment.[53]

Díaz claimed his decision was based on whether Álvarez was willing, or not, to comply with the behavioral and moral standards of the school. The rector claimed that Álvarez' unalterable personality warranted his eviction from the academic community, and not his pardo status. Díaz had already been made aware of Vice President Santander's decree against racial harassment and discrimination in schools. Landa had sent him a copy of Santander's proclamation of January 30, 1826, stating that all necessary measures must be taken to ensure that any Colombian child could be admitted to the Caracas Academy and that he would not experience "harassment or afflictions by indirect means that would obligate them to live outside the school's walls." Landa said that the decree would be fulfilled to the letter and that "Citizen José [Joseph] Álvarez, the vecino of San Carlos who provoked it, would have the support of the Government so that his son will not be bothered in school and so that he will be educated in it as he wishes."[54] Yet the rector continued to contravene Santander's decree in the Álvarez case. His refusal shows that leaders of academic communities demanded some autonomy from government intervention in the administration of their schools. Díaz, for one, believed that he was buttressing the moral integrity of his school by expelling Álvarez.

The Caracas rector used the city's multiracial context as evidence that neither he nor the academy had discriminated against black Colombians. Diaz's claim that a significant number of black men attended his school corresponds to his contemporaries' perceptions of the numerical—albeit not social—dominance of blacks in Colombia's Caribbean communities. Nevertheless, there were marked differences in social standing and obligations between students who attended lectures during the school day and those who boarded at the academy. Álvarez, like the other San Carlos pardos on scholarship whom Díaz mentions, lived among a few dozen other students within the school walls. Strict rules governed student behavior both inside and outside school, which students could leave only in the case of extreme illness, death of a family member, or a citywide religious service. Teachers and, indeed, the entire community, monitored the behavior of boarding students. They represented the school and symbolized the virtues of the city and nation. Day students, on the other hand, while subject to the academy's regulations during classes, were released from those rules as soon as they passed through the school gates. The different freedoms and status of day students and boarding students resulted in quarrels among the students, as well as among the students and other young men in the city who did not attend school at all.

Students may have lived at school, but they were not cloistered there. Male and, in very different ways, female students participated in the neighborhood life around their schools. Adolescent male students, especially, encountered soldiers, apprentices, and un-enrolled children during the three-hour afternoon recess, when they ate lunch, took walks, played ball, or—more worrisome to school administrators—gambled in street games such as dice and cards. In December 1832, Rufino Cuervo, the governor of Cundinamarca, responded to a plea from the director general of education that the provincial government help to restrain the, literally, extramural activities of students.[55] The school director and the mayor testified that elementary students in Bogotá lost "a great part of their time in street games, lounging on porches, and in rakishness before entering the School." The mayor believed that the children's immoderation could be curbed by adding an anteroom to the school, in which a monitor could supervise the students at play before school. The school director argued, too, that "this situation should not be denounced, because even though [he was] not responsible for anything that happens outside of the School, [he had] punished every child that had done wrong." The only other choice the principal had would be to start classes an hour earlier and ask parents to send their children to school exactly at that time. Governor Cuervo agreed to take the "necessary measures to ensure that order should reign among the students."

School administrators worried over extramural activities, but they had little power to control the students. They did, however, find ready support in the government because local officials were just as concerned with public order and morality as were teachers and principals. Parallel problems arose over the signs of academic belonging—articulating which behaviors and clothing identified students as students. School directors and local governments tried to dictate appropriate public behavior—by punishing such misbehavior as gambling when it occurred in school and seeking government aid for misconduct outside it. School administrators also enforced the specificity of school affiliation, primarily through strict codes about school uniforms. Interestingly, these symbols of academic belonging drew both praise and scorn when students mingled with others the street. Several cases in the educational ministry archives involve allegations that nonstudents mistreated students, attacking them either verbally or physically for wearing school uniforms. Although not a large number, the presence of even one such case indicates the seriousness with which the government viewed these conflicts between children and adolescents outside school.

Students quarreled occasionally with other residents during extramural outings. In one case from Pamplona in December 1825, the rector of the school of Nuestra Señora de Las Nieves alleged that soldiers stationed nearby had repeatedly harassed his students.[56] Rector Raymundo Rodríguez charged that the soldiers had been carousing nightly, either in their quarters or the *plaza mayor* (main plaza), both of which neighbored the school. They sang loudly and clanged bells until all hours of the night, disturbing the students' and teachers' sleep. "After a long and barbarous accommodation to this tiresome burlesque," many parents had removed their boys from school. Even though the nightly festivities were terrible, the rector complained most bitterly about the soldiers' treatment of his students during the day. Whenever students passed through the school gates, the soldiers set a dog upon them, making them turn back to the school in fear. The soldiers shouted sarcastic remarks at the boys, especially at those wearing the *havitos talares*—an ankle-length uniform similar to a cassock. The army men used "so many expressions [of abuse] that it can only suggest their perpetual vulgarity."

Vulgarity was awkward, but the attacks were also racist. What most disturbed the rector was that the soldiers had targeted an indigenous boy on scholarship to the school.

> They have outraged Colombia in the Person of an Indian on fellowship to this School, disparaging him for having been admitted to the student body. This outrage is worsened most criminally, since they always throw in his face publicly a defect that the Government, far from seeing it as such, considers a credit to all his excellent efforts. The Government certified him and recommended that I admit him among my students. And this is how they behave, Commander Sir, the defenders of the Equality that the Constitution established? Is this the knowledge that they have of the quality (*calidad*) of the Indians who have sprung from, on all four sides, the first Innocent settlers of these Countries?[57]

Rodríguez was clearly dismayed at seeing his students mistreated, particularly a hardworking and intelligent minority student. But it is important that he used the language of constitutional equality to defend the student's right to be educated. He could write an entire apologia, Rodríguez told the army commander, about the "superior nobility of the Indians over those who insult them." However, he thought the laws favoring indigenous people should be clear enough, since the government had dictated that no less justice or opportunity should be shown to Indians than any other Colombian. The rector asked the commander to make his soldiers stop using "ugly and abominable words." And, since Rodríguez knew that "silence can be more eloquent than any number of expressions," he asked the commander to restrain their "excesses" in every way.

The army commander took seriously the allegation that his troops lacked honor and civic virtue. He immediately defended his soldiers against the charges, challenging the scope of the rector's allegations.[58] He assured the rector that he would punish any soldier who injured or offended a student in the future. Still, conflicts between soldiers and students loomed large. A week later, Rector Rodríguez protested that soldiers continued to use bad language in the presence of his students.[59] Since only a narrow street separated the school and the garrison, when students left school, they met soldiers on guard duty, who still insulted them. The situation had worsened, Rodríguez complained, with the growing number of army volunteers who traipsed up and down Pamplona's streets, setting a bad example for his students. Even church services had been disturbed by "the uproar and shouts that echo through that sacred building." The rector was upset by the soldiers' clamor and coarse language, but he alluded to another concern in this second letter. If the young army volunteers "set a bad example," then Rector Rodríguez considered his students susceptible to it. He appeared to worry that army life might intrigue his students and lead them to abandon their education. The rector intended to quarantine his students not only from the soldiers' immorality, but also from their "infectious liberty."[60]

* * * *

In the early nineteenth century, measuring teacher and student morality called into question the markers of character itself. Clothing, behavior, and speech came under new scrutiny as the national government tried to implement its reformist legislation. The political liberalism of the 1820s introduced policies to transform Colombians into virtuous republican citizens. Liberal elite visions of civic membership were still rooted in older perceptions—and problems—of race and status in this slaveholding society. Enlightenment and early national aspirations to instill civic virtue in the lower orders through secular, state-led primary schools led to vast, but often vague, educational legislation. Efforts to license teachers and standardize school dress, ongoing demands that towns find ways to afford schools despite widespread poverty, and more insistent legislation about nondiscriminatory admissions policies showed the ways in which the young republican government tackled multiple challenges to its state-building projects. The country's first secretary of the interior spent a decade judging local tax matters, the appropriation of indigenous lands, and admissions and scholarship policies and deciding which constituents—girls or boys; rich or poor; white, black, or indigenous—would be granted the opportunity to be educated. His judgments from the 1820s show that while he took seriously the allegations of racial discrimination, he acted infrequently on behalf of black Colombians. The fact that Álvarez's case brought the executive branch to reaffirm the national nondiscrimination policy marked the large extent to which the government tried to bolster liberal ideas throughout the country. At the same time, feeble state support for indigenous education complicated this language of racial equality. Time and again in the first decade of independence, provincial governors failed to establish schools for Indian children, instead allocating communal indigenous revenues to schools for white children.

The Enlightenment ideal of universal education spread throughout early national Colombia. But new voices clamored for their right to education. Local conflicts over racial equality, curricula, and funding in schools had only just begun. Symptomatic of the national government's bind, private school associations sprouted up in the early 1830s. Faced with harsh economic and political conditions, the government education ministry allowed "societies of friends" (*asociaciones de amigos*) to assume responsibilities for maintaining public schools and establishing private ones. These private education societies soon sprang to the forefront of the national school movement. The societies held yearly student examinations, published newsletters and textbooks, debated the merits of foreign curricula, and raised funds for local schools.

Early national reforms to centralize, secularize, and strengthen the primary school system left an uneven legacy in Colombia. By the mid-nineteenth century, the Colombian government had surmounted several problems. By the 1850s, indigenous Colombians could look toward several important legal precedents to assure their continued presence in schools. Yet only few indigenous students were able to attend secondary schools, academies, and universities. Upper-class Afro-Colombians had opened doors to higher education in cities along the Caribbean coast, but deep-rooted racism—and other legacies of the Colombian slave system—undermined that limited success. Many more schools opened their doors to Colombian girls, but few took lessons beyond domestic economy and Bible study. Still, on the whole, the Colombian primary school system expanded from the 1820s onward. Most scholars point to the last quarter of the nineteenth century as a turning point in school development in Colombia, culminating in the massive national system of the early twentieth century. However, this later explosion of systemic public education depended upon the earlier and much slower work of countless teachers, priests, mayors, parents, and town councils to set their districts on a path toward universal education.

The nineteenth-century state shifted many obligations to private hands. Multiple pressures on public resources seriously limited the first decades of school formation. The mid-century marked a contraction of hopes for egalitarian reforms in Colombia, but not their demise. By the mid-1800s, the national education policy had moved toward a pragmatic conservatism. More and more multiracial, nonelite children stood on the doorsteps of schools. At the same time, the Colombian elite held tighter to the reins of their authority to oversee education. By extension, participation and membership in the nation continued to exclude the country's majority—indigenous, black, and poor white Colombians.

Notes

1. Given the largely mountainous or jungle terrain, postmen traveled on foot, or relied on locals commuting for trade to deliver the mail to their small-town neighbors.
2. And rural schools would continue to suffer. Even after decades of reforms, when botanist Isaac F. Holton traveled to Colombia in the mid-nineteenth century, the first school he saw shocked him. In one room of a hut, a "mere boy" taught a dozen others, many unclothed. Girls, he noted, "must generally learn as they can at home, or, as is too often the case, go ignorant." See Isaac F. Holton, *New Granada: Twenty Months in the Andes;* ed. C. Harvey Gardiner (Carbondale: Southern Illinois University Press, 1967), 6–7.
3. Jane M. Rausch, *The Llanos Frontier in Colombian History, 1830–1930* (Albuquerque: University of New Mexico Press, 1993), 25–60.
4. Other teachers succeeded in maintaining or regaining their employment after independence: So, too, did clergymen who taught primary school, such as priest Maria Antonio del Bastos in Foro, friar Josef Miguel Garay in Bogotá, and friar Civilo

Busto, who taught in the Convento of San Francisco in Bogotá. See Archivo General de la Nación [Bogotá, Colombia], Sección República, Fondo Ministerio de la Instrucción Pública (AGN SR FMIP), Tomo 107, folios 732–737; Tomo 108, folios 255, 346–348.

5. For example, confusion arose over which teacher had been hired for Zipaquirá's school. In July 1822, Cundinamarca Intendant Estanislao Vergara named José María Bernal as the new schoolteacher. Mayor Tomás Barriga y Brito had heard, instead, that the noted Lancasterian teacher José María Triana was expected to replace their interim teacher Miguel Lizarralde. The mayor wanted Lizarralde to keep his job. Vergara ignored the mayor, reversed his earlier appointment of Bernal, and sent Triana instead. An infelicitous decision, it turned out, since Triana later referred to the three years he worked in Zipaquirá as "that bitter task" of the "painful and terrible practice" of teaching children. Triana had already held several teaching positions in succession. Just prior, Triana transferred from Tunja's Lancasterian school to replace the famed French teacher Pierre Comettant as the principal teacher in Bogotá's mutual school. Comettant had taught the Lancasterian method to priests. But Triana did not stay long enough in Bogotá to do the same; scant months later, he left for Zipaquirá [AGN SR FMIP, Tomo 108, folios 213, 215, 327, 329, 333–334].

6. AGN SR FMIP, Tomo 109, folios 338–339. Honda, May 19, 1826.

7. AGN SR FMIP, Tomo 109, folio 339.

8. Ibid.

9. AGN SR FMIP, Tomo 109, folio 341. Honda, June 30, 1826.

10. AGN SR FMIP, Tomo 109, folio 341.

11. Ibid.

12. Ibid.

13. AGN SR FMIP, Tomo 109, folio 342. Honda, July 23, 1826.

14. AGN SR FMIP, Tomo 109, folio 342.

15. Ibid.

16. AGN SR FMIP, Tomo 109, folio 341.

17. AGN SR FMIP, Tomo 109, folio 342.

18. AGN SR FMIP, Tomo 109, folios 337 and 343. August 16 and July 24, 1826.

19. Guerra's provenance is not known, but he appears to have been trained in Bogotá.

20. AGN SR FMIP, Tomo 107, folios 613–614, p. 3. Francisco de Paula Santander, "Primary School Regulations for the Province of Antioquia," Bogotá, October 6, 1820.

21. A decade after arbitrating Guerra's case, Restrepo had become the director general of the national school system, and he investigated the provincial school of Guanentá [AGN SR FMIP, Tomo 113, folios 683–718. Guanentá, Socorro Province, March 8, 1838]. The documents indicate that the government had not ratified universal rules of school discipline, even with the national education program designed in 1826.

22. For example, in 1854, a Peruvian court allowed that the whip could be used within the home against servants and family members, but that a teacher could not lash her students. See Sarah C. Chambers, *From Subjects to Citizens: Honor, Gender, Politics in Arequipa, Peru, 1780–1854* (University Park: The Pennsylvania State University Press, 1999), 199.

23. Carlos Newland, "Spanish American Elementary Education before Independence: Continuity and Change in a Colonial Environment," *Itinerario* 15, no. 2 (1991), 84.

24. José María de Mier Riaño, ed., *Gran Colombia. Volumen I: Decretos de la Secretaría de Estado y del Interior, 1821–1824* (Bogotá: Presidencia de la República, 1983), March 11, 1822, 62–63.

25. Riaño, *Gran Colombia*, vol. 1, 62.

26. AGN SR FMIP, Tomo 107, folios 421–422. Tunja, October 30, 1829.

27. Colombian law (December 19, 1828) dispossessed small-property owners who were found to have let their lands lie fallow or underutilized them.

28. The term "salinas" denotes both the place and the activity of salt extraction. However, "salt mine" is a poor translation since the salt extraction process occurred above ground in this Colombian highland region. See Marianne Cardale de Schrimpff, *Las Salinas de Zipaquirá: Su Explotación Indígena* (Bogotá: Fundación de Investigaciones Arqueológicas Nacionales, Banco de la República, 1981). See also the chronicles of Francisco Antonio Moreno y Escandón, *Indios y Mestizos de la Nueva Granada a finales del Siglo XVIII* (Bogotá: Biblioteca del Banco Popular, 1985).

29. Upon independence, the government nationalized and monopolized the salt mines, which provided one of the most important sources of its revenue for several decades. César Mendoza Ramos, *Colombia: Inercias y Cambios (1780–1850)* (Barranquilla: Editorial Antillas, 1992), 33–35; Juan Friede, *Rebelión comunera de 1781: Documentos*, vol. 1 (Bogotá: Colcultura, 1981), 77–100; Pablo Cardenas Acosta, *El movimiento comunal de 1781 en el Nuevo Reino de Granada*, vol. 1 (Bogotá: Tercer Mundo, 1980), 18–29; John Leddy Phelan, *The People and the King: The Comunero Revolution in Colombia, 1781* (Madison: University of Wisconsin Press, 1978); Luis Fernando López Garavito, *Origen y Fundamentos de la Hacienda y la Economía Colombianas* (Bogotá: Universidad Externado de Colombia, 2000), 41.

30. AGN Sección Archivo Anexo, Fondo Instrucción Pública (SAA FIP), Tomo 4, folios 533–536. Zipaquirá, December 6 and 21, 1813.

31. AGN SAA FIP, Tomo 4, folio 536. Zipaquirá, December 21, 1821.

32. AGN SAA FIP, Tomo 4, folio 535. Zipaquirá, December 6, 1821.

33. AGN SAA FIP, Tomo 4, folio 536, marginalia. Bogotá, January 18, 1814.

34. Aníbal Galindo, *Historia Económica i Estadística de la Hacienda Nacional desde la Colonia hasta Nuestros Dias* (Bogotá: Imprenta de Nicolás Ponton i Compañía, 1874). See Cuadros 6 and 7, 111–112.

35. AGN SR FMIP, Tomo 108, folio 361A. Bogotá, April 2, 1823.

36. Vecinos were the tax-paying residents who usually constituted the most powerful local political blocs and who were predominantly white—a racial division that was slow to change in the early national period.

37. AGN SR FMIP, Tomo 108, folio 361B. Zipaquirá, March 10, 1823.

38. AGN SR FMIP, Tomo 108, folio 361B, marginalia. Zipaquirá, March 25, 1823. The salinas' manager reported that the revenue from the salt factories and its lands, including rents from the grazing fields, totaled 550 pesos. The government had been using these revenues to support military operations, but the manager figured that 200 pesos could be redirected to the teacher without much harm to the army.

39. AGN SR FMIP, Tomo 108, folios 381–382. Zipaquirá, May 2, 1823.

40. AGN SR FMIP, Tomo 108, folio 384, 379 marginalia, and 310–312.

41. Nemocón also experienced rapid teacher overturn. AGN SR FMIP, Tomo 108, folios 236–237.

42. AGN SR FMIP, Tomo 108, folios 292–293. Nemocón, October 20, 1823.

43. AGN SR FMIP, Tomo 108, folio 292.

44. AGN SR FMIP, Tomo 108, folio 294. Nemocón, November 10, 1823.

45. AGN SR FMIP, Tomo 108, folios 291–292. Nemocón, October 20, 1823, and Bogotá, November 21, 1823.

46. AGN SR FMIP, Tomo 108, folio 292.

47. AGN SR FMIP, Tomo 113, folios 501–503. Pamplona, December 19, 1837.

48. AGN SR FMIP, Tomo 113, folios 503–504. Pamplona, December 20, 1837.
49. AGN SR FMIP, Tomo 113, folio 504.
50. AGN SR FMIP, Tomo 113, folio 501.
51. AGN SR FMIP, Tomo 123, folio 489. Rector Joseph Nicolás Díaz to provincial Intendant J. Manuel Landa, in response to the declaration made by Citizen Joseph Álvarez to the Chamber of Representatives on February 28, 1826. Caracas, March 13, 1826.
52. AGN SR FMIP, Tomo 123, folio 489.
53. Ibid.
54. AGN SR FMIP, Tomo 123, folio 650. Intendant J. Manuel Landa to Secretary of the Interior Restrepo. Caracas, March 7, 1826.
55. AGN SR FMIP, Tomo 110, folio 387. Bogotá, December 17, 1832.
56. AGN SR FMIP, Tomo 123, folio 136. Colegio de Nuestra Señora de las Nieves, Pamplona, December 2, 1825.
57. AGN SR FMIP, Tomo 123, folio 136.
58. AGN SR FMIP, Tomo 123, folio 132. Cuartel General, Pamplona, December 2, 1825.
59. AGN FMIP, Tomo 123, folios 138–139. Colegio de Nuestra Señora de las Nieves, Pamplona, December 10, 1825.
60. Interior Secretary Restrepo received the battery of complaints and defenses a month later and deemed "no resolution" possible in the case. He archived it and no further complaints were registered against the soldiers quartered in Pamplona. AGN SR FMIP, Tomo 123, folio 139, marginalia. Bogotá, January 9, 1826.

PART II

Politics, Ideology, and Policy

5

Historic Diversity and Equity Policies in Canada

Reva Joshee and Lauri Johnson

This chapter is a historical policy study that examines a particular moment in the development of the Canadian diversity policy web. Starting from the position that policy is the result of a complex set of interactions among several state and nonstate actors, we use the metaphor of the web to examine the relationship between and among policy statements and actions in a particular field.[1] Policy is thus understood as more than a single authoritative text, and the notion of "policy actors" replaces the traditional term "policy makers" as we examine the roles of multiple players in the field. We are interested in understanding the complexity that lies behind the creation and ongoing re-creation of official policies. We believe that at its best, policy can be the result of and catalyst to public dialogue about issues that are of central importance to a society. To this end, our work seeks to find ways of making the policy process more open and democratic. Because of our own backgrounds and commitments, we focus particularly on the study of policies addressing diversity and equity. We have found that the 1940s was a pivotal time in the development of diversity and equity policies in Canada. So, as we focus in this volume on transformations in education, it seems most appropriate to concentrate on this period.

In this chapter, we examine the process that led to the development of the Welland Citizenship Programme, an educational program that was implemented in the province of Ontario in the 1940s. This example allows us to see how different actors and influences came to bear on the educational policy process and laid the foundation for later policies in multicultural and diversity education. It also gives us the opportunity to think in nontraditional ways about the relationship between policy and practice.

Background

From at least the time of the British appropriation of the land now called Canada, official policies have encompassed coexisting discourses of assimilation and respect for diversity. The Royal Proclamation of 1763, for example, confirmed the rights of self-government of the aboriginal peoples of the land.[2] At the same time, it expressed the desire of the British to assimilate the French.[3] Even the Quebec Act of 1774, which sanctioned the continued existence of the French language, culture, and legal system, and the Constitutional Act of 1791, which resulted in the creation of two administrative units, Upper and Lower Canada, and established the separation of British and French Canada, did not deviate from the goal of assimilation.[4] These policies were, instead, a pragmatic response to the fact that at the time, few British people were coming to the region, so rapid assimilation of the French was simply not possible. And although assimilation remained the overriding goal of Canadian policies until at least the 1940s, by 1900 the prime minister of the day, Sir Wilfred Laurier, uttered words that have since been quoted by numerous politicians and ethnic leaders:

> Three years ago when visiting England at the Queen's Jubilee, I had the privilege of visiting one of those marvels of Gothic architecture which the hand of genius, guided by an unerring faith, had made a harmonious whole in which granite, marble, oak and other materials were blended. This cathedral is the image of the Nation that I hope to see Canada become. As long as I live, as long as I have the power to labour in the service of my country, I shall repel the idea of changing the nature of its different elements. I want the marble to remain marble; I want the granite to remain granite; I want the oak to remain oak.[5]

Despite the rhetoric, hardly anyone would argue that any of the early policies or discourses translated into enlightened practice. The First Nations in Canada are still struggling to have their rights to self-government recognized, and for most of the history of Canada, French Canadians have been dominated politically and economically by English Canadians.[6] Furthermore, even the superficial recognition of diversity did not initially translate to other immigrant groups. Canada maintained close ties with Great Britain not only as part of the empire but also as a "North American outpost of Britain";[7] therefore, part of the national project was to fashion an identity based on British ideals. The fact remains, however, that for whatever reasons, recognizing the rights of diverse groups was already part of the policy landscape by the time Canada officially became an independent nation-state in 1867.

From the time of Confederation, education policies were a major tool in addressing diversity, so it is not surprising that we see the coexistence of assimilationist and cultural-pluralist discourses in these policies as well. Because constitutionally, education is a provincial responsibility, we find that educational policies played different and somewhat contradictory roles in different parts of Canada. The overall mission of public education, from its inception in 1847, was to instill patriotism in Canadian youth. Schools were meant to be a homogenizing force that would work with immigrant and native-born children and their families to create "good Canadian citizens" in the image of British loyalists. While residential schools with the express goal of assimilation were established for aboriginal children, it was widely believed that certain other groups could not be assimilated and, therefore, should remain separate. Segregated schools were established for African Canadian children in Nova Scotia and Ontario. Additionally, there were several attempts to segregate Asian Canadian children in British Columbia.[8] At the same time, schools in the prairies were experimenting with education in languages other than English.[9]

The feelings of nationalism engendered by World War I led to an increased emphasis on Canadianization. The bilingual provision was removed from the Manitoba statutes in 1916 and provisions to teach ancestral languages were removed from legislation in Saskatchewan in 1919. By 1921, 15 percent of the population of Alberta and 18 percent of the population of Manitoba and Saskatchewan were newcomers from Central and Eastern Europe, and "there were fears that illiterate immigrants would lower the cultural level of the whole country and undermine British governmental institutions."[10] It was in this climate that J. T. M. Anderson (1918), then inspector of schools in Yorkton, Saskatchewan, wrote his noted book, *The Education of the New-Canadian,* subtitled, A Treatise on Canada's Greatest Educational Problem. Anderson, who later became minister of education and then premier of Saskatchewan, emphasized the urgent need to assimilate newcomers to the Anglo-Canadian culture. Failure to do so was the greatest threat to national unity, and any group that resisted assimilation was to be viewed with fear and suspicion. He supported the work already under way with immigrant children, which he saw as the "paramount factor in racial fusion," but he also advocated an active campaign of adult education. In this campaign he needed to enlist the support of all Anglo-Canadians and so encouraged people, for the good of the nation, to overcome their prejudices about socializing with newcomers.[11]

For a number of reasons, the 1940s saw increased attention being paid to issues of diversity in education. Initially, diversity in education programs formed part of a larger state-sponsored strategy to ensure the support of non-British and non-French Canadians in the war effort. After the war, these programs were recast as part of the new plan to implement the Citizenship Act (1947). Both during and after World War II, community groups, educators, and some public servants used the opportunity created by the rhetoric surrounding these policy initiatives to push for greater commitment to diversity as part of the public agenda. Labor organizations worked alongside religious groups, ethnocultural groups, and civil liberties organizations to protest educational segregation and exclusion, to introduce intercultural education programs to educators, and to lobby for human rights policies. Educators worked with various community groups and government agencies to conduct antidiscrimination seminars, produce curriculum materials, and organize conferences to explore issues of diversity and citizenship. Policy developers from all levels of government were involved in several of these initiatives and took what they learned back to their own work. It was against this backdrop that the Welland Citizenship Programme came into being.

Welland Citizenship Programme

Welland is a small community in southern Ontario, near Niagara Falls and very close to the Canada-United States border. As in many other communities across Canada in the immediate post–World War II era, the people of Welland were awakening to a new sense of national identity separate from Britain and were becoming more aware of cultural diversity as a feature of their community. The task at hand was to work with both the new understandings of national identity and cultural diversity to forge a cohesive democracy. The Citizenship Programme was one avenue through which this was to be accomplished.[12]

The Welland Citizenship Programme was, in fact, an adaptation of the Springfield Plan from Springfield, Massachusetts. The Springfield Plan had been developed in the 1930s in the United States as an intercultural education program, which, although most closely associated with Springfield, was implemented in several communities across the United States.[13] The Springfield Plan was seen as innovative in that it directly addressed intolerance and that it expressly included the larger community as well as schoolchildren in its reach. The Welland Programme targeted the same audience in its discussion of citizenship and tolerance.

The Welland Programme had five principal objectives: (1) the development of the individual; (2) the development of the art of living, learning, working, and thinking together; (3) the development of a community spirit; (4) the development of a national spirit; and (5) the development of a world spirit.[14] Each of these objectives was further defined through several subobjectives focusing on attitudes, skills, and abilities, many of which made reference to addressing prejudice and promoting tolerance.

Significantly, the program examined both racial and class-based intolerance. It included suggested lessons for grades one through eight, with an emphasis on curriculum links to English language arts and social studies but with an acknowledgement that citizenship, as defined by the program, "should permeate the whole school plan."[15] It also included several suggestions for ways to involve students in the life of the community. While the program was only in place for about three years, the process through which it came into being opens a window on the complex way in which various actors and influences came together to attempt to make Canadian education policies and practices more culturally inclusive.

Linking Citizenship and Diversity

As we have noted elsewhere,[16] work done in Canada during World War II linked the notion of cultural diversity with citizenship. When the war ended, the Canadian government immediately began work on the development of a Citizenship Act. Prior to 1947, when the act came into effect, Canadians held British passports and were legally considered British subjects resident in Canada. The act was one of many initiatives designed to develop a sense of a unique Canadian identity, and cultural diversity was seen to be a part of this identity. In his remarks at the close of the debate on the first reading of the Citizenship

Bill in 1946, the minister responsible, Paul Martin, Sr., explicitly linked citizenship and cultural diversity, claiming, "Fortune has placed this country in the position where its people do not all speak the same language and do not all adore God at the same altar. Our task is to mould all these elements into one community without destroying the richness of any of those cultural sources from which many of our people have sprung."[17]

At the same time, however, discourses of exclusion were propagated. In reference to the new immigration regulations that were to be introduced, then prime minister W. L. Mackenzie King noted, "There will, I am sure be general agreement with the view that the people of Canada do not wish, as a result of mass immigration, to make a fundamental alteration in the character of our population. Large-scale immigration from the Orient would change the fundamental composition of the Canadian population."[18] With citizenship and immigration situated in the same department from the late 1940s until the mid-1960s, these conflicting discourses would continue to shape different aspects of Canadian diversity policy for many years.

While the Citizenship Act was still in the draft stages, groups and individuals across the country were already attempting to broaden existing definitions of citizenship and citizenship education. Educational organizations, the major left-of-center political party, faith-based organizations, and some ethnocultural groups wanted explicit inclusion of racial and religious equality. Labor groups supported them in this and also lobbied for attention to the rights of workers. Civil liberties groups focused their attention on the need for guaranteed protection of a range of human rights and fundamental freedoms, including the right to freedom from discrimination and freedom of religion. As part of this work, several groups brought the Springfield Plan to the attention of educators and state-based policy actors.

In 1944, the Civil Liberties Association, a Toronto-based group that came into being during War II for the explicit purpose of defending minority rights,[19] organized two conferences to which they invited representatives of "churches, labour, fraternal organizations, service clubs, school teachers, Home & School clubs, businessmen and others to discuss ways and means of combating racial prejudice and discrimination in Toronto."[20] The Springfield Plan was one of the programs discussed at these two events, and the Civil Liberties Association urged Ontario educators to help bring the program to Canada. In addition, George Tatham, president of the Civil Liberties Association, lectured about and facilitated discussions on the Springfield Plan throughout 1946 at events sponsored by a variety of organizations, including the United Nations Society, the Cooperative Commonwealth Federation (CCF), and the Council on Democratic Action.

Out of the conferences organized by the Civil Liberties Association emerged a coordinating body called the Toronto Committee for Intercultural Relations. The Committee for Intercultural Relations, which included representatives of the Church of England, the African Canadian community, the Civil Liberties Association, the Canadian Welfare Council, the Holy Name Society, the Canadian Jewish Congress (CJC), the Ontario Teachers' Federation, and the Canadian Association for Adult Education (CAAE),[21] was established to coordinate the work of a number of groups interested in intercultural relations and to develop educational projects. It organized several projects that addressed diversity in education in the schools. In the spring of 1945, the committee organized screenings of the film *It Happened in Springfield,* which outlined the work that had been done through the Springfield Plan, as a way to acquaint educators and others interested in intercultural relations with the kinds of programs that were already in place in the United States.[22] By 1947, members of the committee focused an integral part of conferences and public meetings on intercultural relations in Canada.

Also in the mid-1940s, the Jewish Labour Committee (JLC), an organization with branches in the United States and Canada, was becoming increasingly interested in the issue of intercultural education. By 1946, the JLC in Canada had taken up the fight against racial intolerance. Unlike their U.S. counterparts, they decided to work behind the scenes, allowing the Joint Labour Councils Against Racial Intolerance to lead the work in intercultural education within the labor movement. Although Canadian unions had had a history of intolerance, by the 1940s, union leaders, at least, were beginning to see racism as wrong for three main reasons: racism was morally reprehensible, it was a threat to freedom and justice, and it was a threat to union solidarity.[23] The Joint Labour Councils conducted informal education sessions with union members and other groups. They showed films addressing racial tolerance and facilitated discussions. In addition, they lobbied for Fair Employment Practices legislation, supported racialized groups who were combating discriminatory practices, and worked with other groups involved in intercultural education.[24] The Joint Labour Councils also brought information about the Springfield Plan and other intercultural education programs to the attention of Canadian educators and policy developers. *It Happened in Springfield* was among the material Joint Labour Councils used in their education sessions, and information about the Springfield Plan was included in the materials the JLC possessed and distributed.[25]

In 1945–1946, the CJC also highlighted the Springfield Plan as a promising approach to combating prejudice. The CJC worked with the Canadian Citizenship Council (CCC) and the CAAE to make "Probing Our Prejudices" the topic for one of the broadcasts of the Citizens' Forum in 1946. The Citizens' Forum was an innovative civic engagement project of the CAAE and the Canadian Broadcasting Corporation (CBC) in the 1940s. The CAAE had organized hundreds of discussion and listening groups across the country that met once a week from October to April to listen to a radio broadcast on a predetermined topic of national importance. Each group would have received in advance of the broadcast some background material on the topic, a study guide, and a list of questions. After they listened to the broadcast, they had to discuss the issues, and a convener from each group would report on the discussions to the provincial secretary for the forum. The provincial secretaries were then responsible for developing summaries of the reports, which were then broadcast the week after the original broadcast on the topic. By the late 1940s, some of the discussion groups for the Citizens' Forum were located in high schools, including the high school in Welland.

In the spring of 1946, the Citizens' Forum took up the topic of prejudice. The CCC prepared the study guide for the session and included in it information on "how we acquire our prejudices; the myth of race; race intolerance and democracy; the Springfield Plan; parent and teacher education."[26] In addition, a leaflet on the Springfield Plan was included as part of the background material. The material was sent out to a mailing list of 20,000, which included 5,000 school principals and several educational organizations in addition to the established discussions groups.

Additionally, in late 1946 and early 1947, community groups including the Young Men's Hebrew Association (YMHA), Young Women's Hebrew Association (YWHA), CJC, and the United Church sponsored events that brought to Ontario prominent Americans who had been associated with the Springfield Plan. Clyde Miller, widely seen to be the architect of the Springfield Plan, came in October 1946 and gave at least one public lecture that received extensive coverage in the *Toronto Daily Star*.[27] Mary O. Pottenger and Clarence I. Chatto, two educators from Springfield, led a public discussion on the Springfield Plan at the Bathurst United Church in February 1947.

Through one or all of these initiatives, officials of the Department of Education in Ontario became familiar with the Springfield Plan. The government of Ontario, like other governments across North America at the time, was concerned that democratic principles were going to be eroded by communist philosophy.[28] Some government officials saw the Springfield Plan as a potentially important vehicle through which support for Canadian democracy and citizenship could be promoted. While there was considerable pressure to implement the program in Toronto,[29] Toronto schools appeared to be somewhat reluctant to adopt the Springfield Plan. Some officials of the Toronto School Board reported that they were already implementing the principles of the Springfield Plan and, therefore, did not need to adopt the plan itself.[30] The Department of Education supported the development of pilot projects based on the Springfield Plan in three Ontario communities, Welland, Fort Erie, and Kirkland Lake,[31] but only two of the school districts, Welland and Kirkland Lake, actually developed and implemented programs.[32] The choice of schools seems to have been predicated on the interest of local school inspectors, principals, and school board officials. In 1946, the Department of Education provided support for a group of educators to travel to Springfield to learn firsthand about the Springfield Plan and to begin the process of adapting the program for Ontario schools.

While the Welland Programme ceased to exist after 1948, there is some evidence to suggest that the Springfield Plan and its offshoots continued to be important in Canada past this point. The Joint Planning Commission (JPC), a group that included representatives from the CCC, the CAAE, and the Canadian Citizenship Branch of the federal government, established the short-lived Committee on Cultural Relations. The committee prepared a report on intercultural policies and programs that was published in 1948 and provided a snapshot of what was happening in the field at the time. This extensive report (about 200 pages in length), which noted that there were many groups and agencies across several sectors working to establish programs and influence policy, highlighted the Springfield Plan and its Canadian adaptations as exemplary programs.[33]

Understanding the Process

So what do we learn from examining the process associated with the Welland Citizenship Programme? We see that information about the program entered Canada through several avenues at roughly the same time. All of the community-based policy actors who drew attention to the Springfield Plan were concerned with expanding the notion of citizenship to include racial equality. The Springfield Plan was one initiative among many that these groups collectively advocated. Individually, these initiatives could be seen as ways to translate policy into practice. Collectively, the initiatives helped to broaden the definition of citizenship. In 1946, the mandate of the citizenship branch of the federal government, for example, was to provide opportunities for newcomers to Canada to gain the knowledge and skills they would need to become Canadian citizens.[34] By 1951, the branch's mandate was:

> To collect and disseminate information designed to promote greater understanding among the various groups of people in Canada.
> To engage in adult education and liaison work with a view to promoting a greater understanding of our Canadian way of life and strengthening our convictions regarding the principles of democracy.
> To accelerate the acceptance of the newcomer and his integration into Canadian life; to increase the present contribution to it on the part of the newcomers.
> To keep contact with other Departments of the federal Government and with departments of the Provincial Governments, especially their Departments of Education; with municipalities; with the Canadian Broadcasting Corporation, the national Film Board, and the foreign language press; and with national and provincial voluntary organizations.[35]

Similarly, by 1948, the citizenship branch of the Ontario Department of Education, which was established primarily to educate adult immigrants,[36] was promoting "planned study in civics, human relations topics, community, national, and international affairs."[37] It is important to note that in the language of Canadian governments in the 1940s and 1950s, "human relations" was a term used to encompass intercultural and intergroup relations and a variety of activities designed to reduce prejudice.

In this example, we see that the federal government's work in citizenship helped to set the stage for the provincial government's work in citizenship education. Work initiated by a variety of community-based policy actors and supported by the provincial government, in turn, had some influence on the refinement of federal and provincial policies. In other words, even though the Welland Programme ran only for a few years, it had a more lasting effect in the ways that it, along with other initiatives, helped to shape the policy dialogue.

With additional attention to other work that was going on at the time, we can also determine which actors were not as well represented in the policy process. The National Japanese Canadian Citizens' Association (NJCCA), for example, which was involved with the Civil Liberties Association, was concerned with addressing prejudice in the majority community, but saw the key educational issue as eradication of barriers to equal access to education.[38] African Canadian groups, which were working closely with the Joint Labour Councils, meanwhile, were interested in eliminating segregation in all aspects of society, including schooling.[39] These issues were rarely, if ever, raised as part of the larger discussions around citizenship education.

This disregard may be simply because the majority of policy actors, both community and state-based, were members of the dominant European ethnic groups. They may not have seen the concerns of these two racialized groups as priorities. Another possible explanation for the apparent lack of attention to the concerns of Japanese Canadian and African Canadian policy actors may be found in the politics of the "red scare." By the late 1940s, left-leaning organizations of all kinds, including the Civil Liberties Association, CAAE, unions, and the CCF, were under suspicion of being sympathetic to communism. Moreover, "in the Cold War era, almost any attempt by blacks to deal with racism and prejudice could be dismissed as the product of Communist agitators."[40] As a result, the policy actors may have deliberately avoided what were perceived to be more controversial issues. Whatever the reason, we see from this example that even in conditions that seem open to the participation of diverse policy actors, the priorities of minoritized groups can easily be marginalized.

Traditional approaches present policy development as a linear process based on a series of rational decisions deliberately made by a small group of policy developers.[41] Even the less-positivistic models of policy development that are gaining attention in the field tend to separate policy development and policy implementation on the assumption that policies are relatively independent of each other.[42] This examination shows that real-life policymaking is much more complex. We have long known that the adoption of official policy does not lead to immediate and irrevocable change, yet we have had difficulty defining how to understand the relationship between policy and practice. While we do not claim to have found the definitive answer to this puzzle, we believe that a policy-web approach can help us to better understand the intricacies of the policy process at an important time of educational transformation in Canada.

Notes

1. Reva Joshee and Lauri Johnson, "Multicultural Education in the United States and Canada: The Importance of National Policies," in International Handbook of Educational Policy, ed. N. Bascia, A. Cumming, A. Datnow, K. Leithwood, and D. Livingstone (Dortrecht: Kluwer Publishing, 2005), 53–74.
2. M. Ignatieff, The Rights Revolution (Toronto: Anansi Press, 2000).
3. N. Kelley, N., and M. Trebilcock, The Making of the Mosaic: A History of Canadian Immigration Policy (Toronto: University of Toronto Press, 1998).

4. National Archives of Canada (NAC), MG 31 E55, Walter Tarnopolsky Fonds, vol. 47; Arthur Lermer, "The Evolution of Canadian Policy towards Cultural Pluralism" (CJC, June 1955).

5. NAC, MG 31 E55, vol. 47; Sir Wilfred Laurier cited in Lermer, "Evolution of Canadian Policy."

6. Ignatieff, *Rights Revolution.*

7. R. Joshee, *Federal Policies on Cultural Diversity and Education, 1940–1971* (PhD diss., University of British Columbia, Vancouver, 1995).

8. J. Burnet, and H. Palmer, *Coming Canadians: An Introduction to the History of Canada's Peoples* (Toronto: McClelland and Stewart, 1989); J. W. S. G. Walker, *Race, Rights and the Law in the Supreme Court of Canada* (Waterloo, ON: Osgoode Society and Wilfred Laurier University Press, 1997).

9. Burnet and Palmer, *Coming Canadians.*

10. Ibid., 112.

11. J. T. M. Anderson, *The Education of the New Canadian: A Treatise on Canada's Greatest Educational Problem.* (Toronto: J. M. Dent, 1918), 89.

12. Welland Public Library. *A Programme of Citizenship,* Experimental Ed. (Welland Public Schools, Welland, Ontario, 1946).

13. L. Johnson, "One Community's Total War against prejudice": The Springfield Plan Revisited, *Theory and Research in Social Education* (forthcoming).

14. Welland Public Library. *Programme of Citizenship.*

15. Ibid., 69.

16. Joshee and Johnson, "Multicultural Education," 53–74; R. Joshee, "Citizenship and Multicultural Education in Canada: From Assimilation to Social Cohesion," in *Diversity and Citizenship Education: Global Perspectives,* ed. J. A. Banks (San Francisco: Jossey-Bass, 2004), 127–156; R. Joshee, "An Historical Approach to Understanding Canadian Multicultural Policy," in, *Multicultural Education in a Changing Global Economy: Canada and the Netherlands,* ed. T. Wotherspoon and P. Jungbluth (New York: Waxmann Munster, 1995), 23–24.

17. NAC, MG 28 I 179, vol. 7, File 25–8d, House of Commons Debates, Official Report, Canadian Citizenship, Hon. Paul Martin, Secretary of State and Member for Essex East, Addresses on the Act Defining Status of Canadian Citizenship and Canadian Nationality, 1946.

18. Cited in A. G. Green and D. A. Green, "The Economic Goals of Canada's Immigration Policy: Past & Present," *Canadian Public Policy* 25, no. 4 (1999): 425–451.

19. C. Patrias and R. Frager, "This Is Our Country, These Are Our Rights": Minorities and the Origins of Ontario's Human Rights Campaigns," *Canadian Historical Review* 82, no. 1 (2001): 1–35.

20. NAC, MG 31 K9, Frank and Libby Park Fonds, vol. 8, Letter to the members of the Civil Liberties Association, Past and Present, June 1946.

21. Archives of Ontario. F 1205 Series A1, Tiles of the Canadian Association for Adult Education, Director's files, 1935–1943, a I 'Box, File Inter-cultural Relations, Toronto Committee.

22. NAC, MG28 I28, Files of the Canadian Citizenship Council, vol. 55, An Inventory of Intercultural Programs and Policies in Canada, published by the JPC, 1948.

23. R. Lambertson, "'The Dresden Story': Racism, Human Rights and the Jewish Labour Committee of Canada," *Labour /Le travail* 47(2001), www.historycooperative.org/journals/llt/47/03lamber.html (accessed March 15, 2005); P. Sugiman "Privilege & Oppression: The Configuration of Race, Gender, &

Class in Southern Ontario Auto Plants, 1939 to 1949," *Labour/Le travail* 47 (2001), www.historycooperative.org/journals/llt/47/04sugima.html (accessed March 15, 2005).

24. NAC, MG28 V75, Files of the Jewish Labour Committee, vol. 34, Address by Kalmen Kaplansky, National Director, Jewish Labour Committee of Canada, delivered on CKEY Radio, Toronto, Ontario, July 19, 1948.

25. NAC, MG28 V75, Files of the Jewish Labour Committee, vol. 35, Pamphlets, Conferences on Human Rights.

26. NAC, MG31 E55, Walter Tarnopolsky Fonds, vol. 41, M. Saalheimer, "Prejudice and Canadian Unity" (CJC, July 1946).

27. "Would Immunize People against Race Prejudice," *Toronto Daily Star.* October 28, 1946.

28. Bob Davis. *Whatever Happened to High School History? Burying the Memory of Youth, Ontario: 1945–1995.* (Toronto: James Lorimar, 1995), 27.

29. "The 'Springfield Plan' in Toronto." Editorial, *Toronto Daily Star,* January 11, 1946.

30. "Springfield Plan Old Here in Teaching Racial Amity," *Toronto Daily Star,* January 5, 1946.

31. NAC, MG28 I28, Files of the Canadian Citizenship Council, vol. 55, An Inventory of Intercultural Programs and Policies in Canada, 1948.

32. Davis, *High School History.*

33. NAC, MG28 I28, Files of the Canadian Citizenship Council, vol. 55, An Inventory of Intercultural Programs and Policies in Canada, 1948.

34. NAC, MG28 I179, Files of the Canadian Citizenship Council, vol. 25, Publications and Programmes, 1946, Canadian Citizenship Branch.

35. NAC, RG 26, Files of the Canadian Citizenship Branch, vol. 67, File 2–18–1. Report of the First Annual Conference of the Canadian Citizenship Branch, August 20–25, 1951.

36. "To Give Instructions on Citizen's Duties," *Toronto Daily Star.* October 25, 1945.

37. Archives of Ontario, RG 2–74–3, Files of the Community Programs Branch, Department of Education, Box 1, File on Fort William Community Activities, October 1946 to December 1948.

38. NAC, MG31E55, Walter Tarnopolsky Fonds, vol. 47, Submission of the National Japanese Canadian Citizens' Association to the Special Senate Committee on Human Rights and Fundamental Freedoms, May 1950.

39. Patrias and Frager, "'This is Our Country.'"

40. R. Lambertson, "The Black, Brown, White & Red Blues: The Beating of Clarence Clemons," *Canadian Historical Review* 85 (2004): 755–776. The quote is on page. 770.

41. B. W. Hogwood and L. A. Gunn, *Policy Analysis for the Real World* (Toronto: Oxford University Press, 1990).

42. See, for example, E. Roe, *Narrative Policy Analysis, Theory and Procedure.* (Durham, NC: Duke University Press, 1994); S. Stein, *The Culture of Education Policy.* (New York: Teacher's College Press, 2004).

6

Conference Litmus: The Development of a Conference and Policy Culture in the Interwar Period with Special Reference to the New Education Fellowship and British Colonial Education in Southern Africa[1]

Peter Kallaway

There is a great need for a historical understanding of policy environment: the political and ideological context from which policy, both as rhetoric and reality, arises. The purpose of this chapter is to promote a more nuanced understanding of the transformation of educational discourse in the interwar era as reflected in the deliberations of major professional conferences and to attempt to develop a deeper understanding of how these debates helped shape the background to shifts in British colonial education in the 1930s. The following discussion links professional educational networks promoting the exchange of ideas in the era of New Education and tertiary teacher education development to the emergence of missionary and government "policy" and to the origins of educational studies as a scientific or research field at this time.

In the context of the Depression and the challenge to the principles of the League of Nations posed by the rise of totalitarianism in Germany, Italy, Japan, and the USSR, there is a clear shift in emphasis at the conferences of the New Education Fellowship (NEF) from a pedagogy of personal and individual development associated with the Progressive Era to a hard-nosed appraisal of policies that promote economic growth and social development in a democratic context. This shift is tracked through a review of the key themes of the conferences of the NEF, the International Missionary Council (IMC), U.S. foundations, as well as British Colonial Office policy and other significant networks of educational policy debate.

Central to the interpretation of these events is the question of how far these networks can be said to overlap and complement each other. Are the debates about colonial education sui generis, to be understood in terms of colonial exceptionalism via such notions as "cultural imperialism" or "racist colonial domination," or are they simply the local variant of international policy developments? Are the emergent NEF guidelines for educational policy relevant only to the interpretation of policy developments in Europe, North America, and the Commonwealth Dominions, or are they a central feature for understanding colonial education in the 1930s? To put it the other way around, to what extent does an investigation of developments in colonial education help to throw light on aspects of educational reform in the imperial heartland?

There is a lack of research on the linkages between conferences held under the auspices of the World Federation of Educational Associations (WFEA) and the NEF networks of Europe and North America, and the developments in colonial education/mission education from the time of the milestone World Missionary Conference in Edinburgh in 1910. This is despite the fact that these networks of educational discussion take place in parallel between 1910 and 1940.

This chapter attempts to monitor the emergence of alternative voices at such international conferences by the mid-1930s in the context of the rise of political opposition to imperialism following World War I and the establishment of the League of Nations. It also charts the place and role of South African participation in these events, linked to a wider project to locate the history of education in that region within a revised historiographical framework.

The 1934 South African Education Conference as a Benchmark of Changing Educational Discourse

This chapter focuses on "the new turn" in education associated with the changing focus of the NEF conferences in the 1930s. It examines how shifts in conference foci reflect wider forces that were impacting on many spheres of educational debate at that time. It concerns the changing focus of attention from the educational goals of individual psychological growth and development associated with the progressive movement of the 1920s (strongly influenced by Jung, Freud, and Piaget) to the more politically, economically, and socially located goals and critiques of educational policy and practice characteristic of the 1930s.

Although the dominance of psychology and pedagogy remained the major feature of much European educational debate into the 1930s as demonstrated in the work of the Institut J. J. Rousseau in Geneva,[2] this article seeks to locate the shift to more socially located methodologies that emerged at places such as the new Institute of Education at London University.[3] Although there was still a strong influence of Progressivism and psychology here as well, the overall influence of Bertrand Russell, Alfred Whitehead, Percy Nunn, and Fred Clarke meant that there was a new emphasis on sociology, philosophy, history of education, and comparative education, which was to have significant, long-term implications for the study of education.[4]

These conferences—the international South African Education Conference held in association with the NEF's in 1934 and the NEF conferences in Nice (1932) and in Cheltenham (1936)—were significant milestones in this change in the culture of educational debate and research. Together they represented the emergence of an alternative tradition in educational scholarship and thinking that emphasized the social aspects of education. The 1934 conference was also the first to use the inclusive ethos of the NEF to promote the participation of colonial and missionary educators and a number of other significant academic and political figures associated with the emergent study of African education. It also allowed for the participation of a small number of African delegates for the first time, some of them prominent political figures, and it highlighted the racial problems of South Africa as manifested in the relations between Dutch and English colonizers, and between settlers and the indigenous people. Although these influences did not have a major short-term impact on conference deliberations before World War II, they set the tone for much academic research and policy development after 1945.

The themes of the NEF conferences in the 1930s—"Education and Changing Society" (Nice, 1932), "Educational Adaptations in a Changing Society" (Cape Town and Johannesburg, 1934), "The Educational Foundations for Freedom in a Free Community" (Cheltenham, 1936), and "Education for Complete Living" (Australia, 1937)—demonstrate a significant shift from the themes of the 1920s. These placed more emphasis on childhood, pedagogy, and psychology with themes such as "The Creative Self Expression of the Child" (Calais, 1921), "Education for Creative Service" (Montereux, 1923), "The Meaning of Freedom in Education" (Locarno, 1927), and "The New Psychology and the Curriculum" (Elsinore, 1929).

The South African conference was a truly international event. It took place in Cape Town and Johannesburg in July 1934 and was attended by no less than 4,000 delegates.[5] The Cheltenham conference, while significantly smaller and less representative than the earlier NEF gatherings, had an innovative tone that linked it in some ways to the ethos of the South African conference as distinguished from the earlier European events.

In keeping with the general tone of the NEF, the presenters at these conferences were not exclusively educational experts and academics. In particular, the South African conference included many presentations by administrators, teachers, educational activists and experimenters, and practitioners of many kinds. It exhibited many of the tensions between traditionalists and reformers. In general, these deliberations in South Africa revealed the limits of the New Education in defending democratic rights and seeking to link peace and education in a world increasingly threatened by war. In particular, they were a demonstration of the effects of the increasingly secularized and scientific nature of educational discourse by the 1930s. There was a notable influence from the United States, where the study of education and teacher training was increasingly accepted as a university responsibility. The conferences attempted, obliquely perhaps, to engage with the great political changes associated with the Depression and the rise of totalitarianism, while beginning to explore the educational implications of the changes in European imperialism.

Most significantly for my purposes, the conferences in South Africa and Australia represented the first occasions that such educational gatherings, hosted by an international educational association, had been held outside Europe or North America. Their tone needs to be understood in the context of the passing of the Statute of Westminster in 1931, which allowed the new dominions to stake out a role in the great political transition in the British colonial empire, while participation by Indians and Africans remained marginal. At the South African event, pride of place was granted to analyzing and understanding education in the African colonial context. A range of experts from a variety of disciplines and educational environments, many from academic or professional fields outside the normal ambit of educational studies, were called upon to contribute. At Cheltenham and in Australia, in a rather different atmosphere dominated by threats of war, some of these conference themes were carried forward, but there was less interdisciplinary focus and little participation by delegates from outside the academy.

Educational studies seemed to be becoming more comprehensive and multi-disciplinary just when the NEF was losing much of the unique activist energy in relation to transforming the classroom and practicing pedagogy it had displayed in the previous decade. At the same time, as Fuchs points out, "Looking at these congresses it becomes evident that the internationalization of education in the first half of the twentieth century was a (dominantly) Western enterprise. The concept of scientific education as well as the general ideas of the New Education, such as work schools and the child-study movement, were based on the European concept of Enlightenment and embedded in a socio-political context that differentiated between a 'civilized Europe' and a 'barbaric rest.'"[6] In retrospect, it is also significant that so little attention was given to the great educational experiments taking place in the USSR.

Until the South African event, the challenges of education in the colonial and African contexts had been neglected at NEF conferences, though these had been widely debated in missionary and colonial government circles since the 1920s. These ideas came to be highlighted for the first time in formal scientific and academic educational circles at the NEF conferences and other meetings of educators in the 1930s, particularly at those hosted by the great Christian missionary organizations of the time such as the IMC.

The moving spirit behind the South African conference was Ernst G. Malherbe. An Afrikaner of Huguenot descent who grew up in the Orange Free State, Malherbe graduated from Teachers College in 1924, wrote a major book on the history of South African education,[7] and in 1934 became the director of the newly established South African Bureau of Educational and Social Research,[8] which was initially jointly funded by the Carnegie Corporation and the South African government. The conference was addressed by General J. C. Smuts, minister of justice and deputy prime minister of the *Fusion* government, and J. H. Hofmeyr, the minister of education.[9] Many significant contemporary educationalists were present, including J. J. van der Leeuw, Beatrice Ensor of the NEF, and Professors John Dewey, Pierre Bovet, William Boyd, Fred Clarke, A. V. Murray, and Harold Rugg. Out of nearly 200 papers presented at the conference,

27 were by international academics and 24 by South African academics. These elites of the emergent "scientific" field of educational studies were balanced by a much larger and more varied contingent of educational practitioners from the new educational bureaucracies and the missionary education network in Africa. These practitioners comprised teachers, teacher educators, psychologists, "native administrators," social workers, prison officials, medical experts and, significantly, a few, mostly South African, representatives of the colonized people.

The Development of Professional Educational Networks from the Late Nineteenth Century

The NEF conference in South Africa was the culmination of over half a century of "scientific" debate in education. Eckhardt Fuchs traces the advent of formal educational conferences back to the scientific gatherings that came to be associated with world fairs from the late nineteenth century. He notes that these educational gatherings "covered a wide range of topics from primary education to secondary and higher education, adult education, special schools and school hygiene." In the early days, the delegates represented governments, and "in the context of national exhibits these congresses were organized by the host governments of the international exhibitions and mainly used to introduce various aspects of the national systems to foreign countries."[10] These early conferences were called before the growth of an established conference network linked to educational associations of various kinds. After 1919, many congresses promoted the agendas of the League of Nations in relation to peace education, education for democracy, and human rights. It is important to see the work of these conferences in the context of the general growth of professional debate and research development associated with the rise of the modern scientific academy and as part of the professional ethos of modern education.

The New Education Fellowship in the Interwar Period

The NEF played a central role in the development of a politically and morally based model of internationalism in education. It had its origins in the International Movement in Progressive Education established in the early 1920s, with a "strong commitment among the leaders of the NEF to the fostering of international understanding and a world consciousness through education with support for the League of Nations."[11] The NEF was founded in 1915 "as a rallying point for people of all countries who felt that a radical form of education, based on a proper understanding of childhood and of the unity in diversity of mankind, was essential if ever world peace was to be assured."[12] These developments took place in parallel with the formation of the American Progressive Education Association in 1919.

Kevin Brehony argues that the NEF was a *social movement* rather than a professional or academic organization given that it embraced a range of professional, social, and political agendas. As a social movement, it made key contributions to education by promoting the link between provision and research, by consolidating links between the New Education and the U.S. foundations, and even initiating contacts between the NEF network and those debating and formulating educational policy in the British Commonwealth. By organizing a sequence of conferences that helped to develop the field of educational studies, it provided the context for educational debate while erecting the scaffolding for research and professional development of education in the academy. By the 1930s, the NEF was to make a contribution to educational studies far beyond the scope of its original, activist, organizers.[13] The progressive ethos of the movement was not very scientific/academic in the 1920s. Brehony points significantly to the essentially elitist nature of the movement, not linked directly to struggle for mass education, worker education, or colonial education, though it came to serve all of these ends in various ways during the following two decades. As Percy Nunn put it, "In its origin the NEF was a gesture of revolt against the older tradition and expressed a felt need for reform."[14] The NEF published a magazine, *Education for the New Era: An International Quarterly Journal for the Promotion of Reconstruction in Education,* later renamed the *New Era in Home and School.* The goal of the organization and the journal was to support the outlook and perspective associated with the newly formed League of Nations, which was developing political, social, and economic functions but had no educational functions. The aim of the journal was to promote international dialogue in education, support "the growth of experimental education," and "promote an international fellowship of teachers."[15]

In broad terms, in "this conception of education, the essential thing was not the subjects nor the methods of learning, but right relations between parent and child, and between teacher and pupil."[16] This provided an umbrella for educators, with strong representation by socialists, pacifists, and theosophists, who initially provided the energy and enthusiasm for the project. It was also to be the core of later problems in relation to the coherence of the NEF.

Interwar NEF Conferences and the Links with the British Commonwealth[17]

The majority of delegates attending the NEF conferences came from 18 European countries and the United States. There was also a degree of participation from Japan and China. Later, there were also delegates from India, Pakistan, South Africa, Australia, New Zealand, and Latin America. To further promote the internationalization of the NEF in the late 1920s, significant members like Ovide Decroly gave a lecture tour in Columbia and Adolphe Ferrier to Austria, Rumania, and Turkey. Stewart claims that these visits had a significant influence on policy development and research in these countries.[18]

Focus on research and the development of disciplinary fields were first substantially on the agenda at the 1931 Commonwealth Education Conference in London. The significance of this event, as with the South African conference, was that the organizers managed to include delegates involved in the realities of policy development in India and Africa from government, missionary, and independent perspectives. At the South African conference, there was for the first time a range of speakers who represented other fields of academic inquiry (anthropology, economics, race and cultural studies) and also a variety of administrators and missionary educators who highlighted key aspects of colonial education. These events also highlighted, for the first time, the differences between policy development in the European, dominion, and "imperial dependency" contexts. Although these trends can also be noted at the Cheltenham conference in 1936, it seems that many of the innovations were again muted by the time the Australian conference took place in 1937, when a small number of "international experts" (21 in all) presented papers at the conference to the exclusion of other voices. The precise background to the politics behind these issues is beyond the scope of this chapter.

British Commonwealth Education Conference, 1931

After the NEF conference at Elsinore in 1929, Percy Nunn of the London Day Training School organized a British Commonwealth Education Conference in association with the NEF.[19] This seems to have linked the work of the NEF to his initiative to establish London University's Institute of Education, thus challenging the dominant position of Teachers College and the Institute J. J. Rousseau in the field of educational studies. It was both a strategic move to shape the work of the NEF and an attempt at a more direct intervention in favor of the League's work for democratic education internationally.[20]

Engagement with the Commonwealth education network was an extremely unusual move for the NEF as it meant that for the first time it was associated with policies and practices directly linked to government policy—in particular with regard to the British Colonial Office, the India Office, and, more tentatively, with missionary education and philanthropic movements through the IMC networks operating in the colonial context. It was also a demonstration of the growing political and strategic importance of education and educational policy discourse in the international context as influenced by the League of Nations.

Although there is no space here to explore the details, it is significant that this event linked the activities of the NEF for the first time to a different tradition of educational gatherings associated with the British Colonial Office that had been convened periodically since the beginning of the century (see table 1 below). In turn, the Colonial Office and its subcommittee, the Advisory Committee on Native Education in Tropical Areas (ACNETA), subsequently called the Advisory Committee on Education in the Colonies (ACEC), were closely associated with the missionary education network. These missionaries had also convened a number of large gatherings of international educationalists, especially

Table 1 Imperial and Commonwealth Education Conferences

1902	Conference on Colonial Education
1911	Imperial Education Conference
1925	Imperial Education Conference
1927	Imperial Education Conference
1931	British Commonwealth Conference on Education
1952	Conference on African Education
1968	Commonwealth Education Conference

since the great World Missionary Conference held in Edinburgh in 1910. In many ways the educational concerns of the NEF outlined above were proving significant in shaping educational debate and discourse in these contexts by the 1930s, binding a wide variety of individuals and associations from different contexts into a greater common professional and research identity even if there was little overlap of individuals attending the respective events.

When some 80 invited delegates gathered in London in July 1931 to attend over 60 conference sessions on key British Commonwealth educational issues, they did so in the context of the Great Depression (which prevented some of those invited from attending) and the rise of totalitarianism in Europe and East Asia.[21] They also gathered in the context of the crucial changes associated with the Statute of Westminster and the opening of the Round Table Conferences in India, which signified a major shift in British policy toward the empire.[22]

The tense debate at the beginning of the 1931 meeting regarding the use of the terms "Imperial" or "Commonwealth" captured the mood.[23] At the same time, if change was in the air, it was focused on the developing independence of the dominions or "white colonies" of settlement—Canada, Australia, New Zealand, and South Africa—rather than on the concerns of the small number of apparently frustrated Indian delegates who attended and often led the discussion in the program on "Problems of India." There were no Africans present, though there were many delegates with significant experience in the field of colonial education. Although the conference was held under the auspices of the NEF, it did not follow the usual protocol, as the delegates seem to have attended by invitation rather than by virtue of their membership of the traditional NEF network.[24]

Malherbe emphasized that the Statute of Westminster could be seen as part of an effort to give legal definition to the political consequences of "a new type of leadership, one no longer wedded to the concept of dominance but based upon the ideal of sympathetic guidance which gives full recognition to (national) individuality" that stressed "government not as an exercise of power, but as an agency of service."[25] He argued that the aims and purposes of education in the Commonwealth could be reduced to the unifying ideals for a new education of which the essential point was the abandonment of the idea of cultural dominance, which formed the cement of the old British Empire. Within the ambit of the unifying scientific spirit that transcends creed and culture, he argued, "We may define this ideal as a belief in the value of diversity and the desire for the full development of the culture of each individual group." In emphasizing the right and need to develop cultures other than that of the imperial power, Malherbe was of course emphasizing the rights of Afrikaners (or French Canadians) to language and culture in the Commonwealth. It is unlikely that he saw the irony that is so obvious to us in hindsight. The very rights and freedoms that he was claming as part of the Commonwealth pact were to be the stuff of the Asian and African nationalist revolutions of the future. They must have been only too obvious to the Indian delegates.

Responding to Malherbe, Percy Nunn identified the reactions of those who supported policies of white separation and white integration. He also pointed to the increasing political importance of the relationship between whites (colonists) and colonial peoples.[26] In particular, he raised the "enormously important question" of the attitude of the Commonwealth community as defined in the Statue of Westminster "towards the indigenous peoples of Africa." Although the Colonial Office had attempted to set guidelines for colonial education since the early 1920s, it gradually became clear that the much vaunted policy of *adaptation* had its problems. The attempt to avoid a rigid European curriculum in African schools, and to embrace aspects of indigenous education along with the principles of rural education developed for blacks in the United States were well meaning but were often rejected by Africans as paternalistic and a recipe for inferior education. Nunn saw commitment to an education that attempted "to help him build up a characteristic African individuality of his own" as having positive aspects, but in places such as South Africa such a policy of differentiation risked permanently dividing the traditional English curriculum, the Afrikaans/Dutch stream, and specifically African education. This reflected many of the concerns so lucidly indicated by Victor Murray in his critique of the Phelps-Stokes Commission's findings and the subsequent Colonial Office report *Education Policy in British Tropical Africa* (1925).[27]

It was not just these issues of African colonial education that influenced the tone of the gathering. The presence of a number of Indian delegates, side by side with high-ranking British Indian administrators, was entirely new to NEF forums, and had great significance for the tone of the conference. The point could not have been lost that the white colonies of settlement—the dominions—were being granted equal political status with the "mother country," while Indian negotiations for self-government were moving at a snail's pace. Indeed, Professor Shahani from India pointed out that movements for international and interracial understanding tended "to confine themselves to Europe, so that Asia knows nothing of them." The internationalism of the League seldom had anything to say about India and other parts of Asia, and by implication the colonial context in general.[28]

Reforms in Indian education were being shaped despite the effects of the Depression. Sir Philip Hartog had been chair of the Education Committee of the Indian Statutory Commission 1927–1930 (Simons Commission), which had been appointed to report on the system of government in India and make proposals for reforms.[29] Issues of provision, language, and culture; medium of instruction; religion; "the proper balance between purely literary education and technical/vocational education"; girl's education; rural education; adult education; and higher education were discussed at the conference. A major feature of the debate was the religious issue and how mass schooling of Muslims, Hindus, and Christians was to be arranged in the same schools. There was an intense debate about the nature of the "national education" that would be required for an independent India. At the same time there was a degree of disquiet about the ways in which "national education" was being defined, as the most vocal groupings and prosperous elites tended to support the traditional colonial curriculum. All of this may help explain the extreme caution with which the British tackled these issues in an era of volatile politics[30] and may throw light on the influence of Indian education reforms on later initiatives in Africa and elsewhere.

It is also notable that this conference took place in the context of the publication of the pivotal report of the IMC, *The Remaking of Man in Africa*.[31] This amounted to a comprehensive report on the state of missionary education in Africa and a summary of work being carried out by the African Education Group of the IMC and Conference of British Mission Societies (CBMS) network under the guidance of J.H. Oldham at Edinburgh House, the headquarters of the IMC.[32] The report aimed to set out a strategy for the new missionary era at a time when the Colonial Office was seeking innovative thinking in the area.[33] In many ways it is possible to see such contributions in the context of emergent debate about what was identified as "problems of adjustment" in education. The task for the government was now "not merely to supervise and secure efficiency on the secular side of missionary education but to subsidise it, and provide direct educational institutions to supplement the mission activity which is still being carried out by Christianity and Islam."[34]

The tone of the conference is of some significance. It seems to have managed to fruitfully bridge the gap, at a key moment of political change, between the inward-looking European notion of educational reform in terms of personal development, and wider visions of such reform as an aspect of democratic governance, nation –building, and modernization in a dominion and colonial context.

NEF in South Africa, 1934

After promising signs of change in educational debate at the 1932 conference in Nice, the South African event in 1934 proved to be of great significance in focusing attention on the social, economic, and political aspects of education. Its theme, "Educational Adaptations to a Changing Society," reflects that emphasis. Malherbe's general affirmation that the principles of the NEF were essentially about addressing "problems of human relationships" demonstrates something of the earlier nonpolitical ethos of the organization. The conference provided delegates with an opportunity to engage not only with the consequences of the Depression and the prospects for democracy in Europe and Asia, but also, for the NEF delegates, with educational issues in the unfamiliar context of colonial Africa.

In terms of Malherbe's local agendas, it is clear that hosting the conference strengthened his position very considerably as director of the South African Bureau of Educational and Social Research. It came at a strategic time when there was a significant move to the right in South African politics with the founding of the *Fusion* (coalition) government in 1933 under General Hertzog and the rise of the fascist-aligned Purified National Party (the Hersigte Nationale Party) under D. F. Malan. The Native Economic Commission had recently reported on "Native Education" and recommended an extreme form of racial segregation, education, and employment.[35]

Richard Glotzer provides evidence of support to Malherbe from Teachers College and the Carnegie Corporation in what they considered his important work in South Africa at a time when he, like Charles Loram, was considering emigration.[36] A generous grant from the Carnegie Corporation of New York had been allocated to Malherbe for his study of the poor white problem in South Africa,[37] and another grant was forthcoming to enable a number of prominent international educationalists to attend the conference.[38] Further grant monies were made available from various educational institutions and state education departments.

The South African event engaged with a specific set of issues relating to education outside the conventional international and NEF framework by setting aside substantial time—for 48 papers from a total of over 300—for debates on "Education in a Changing African Society."[39] This aspect of the conference was organized by the prominent South African liberal J. D. Rheinallt Jones, who was closely associated with the Carnegie-funded South African Institute of Race Relations (SAIRR).[40]

In opening the deliberations, Rheinallt Jones invited consideration of "how far the system is meeting the needs of the African child today" and asked for an examination of the changes in African society, "their causes and problems arising, and a critical assessment of education as an effective instrument for the adjustment of the African child within its changing world."[41] In keeping with the ethos of the NEF, many of the participants in this section were educational practitioners of various kinds working in different contexts, but the overall emphasis was on the education of Africans, significant in itself for an international conference meeting in a context where the politics of a racially divided South Africa had moved sharply to the right.

Another novel and significant phenomenon was the presence of a small number of African delegates. This may have set the tone for further African participation at the NEF Cheltenham conference two years later. Dr A. B. Xuma, an American-trained medical practitioner in Johannesburg, prominent member of the Joint Council of Europeans and Natives, and African nationalist, spoke on health and diet change in urban African areas; Don Jabavu, South African Native College, Fort Hare, spoke about the nature of children arriving at the school in South Africa; Don M'Timkulu of Healdtown College and Rev. K. T. Motsete, Tati Training Institution, Bechuanaland, raised questions about language; and E. B. Mahuma Morake, principal of Wilberforce Native Training Institution, Evaton, Transvaal, spoke on the education of girls in a joint presentation with her mentor, Mabel Carney of Teachers College, in "the only joint presentation that crossed racial lines."[42]

One speaker who set the international tone of the conference was Dr. Gustav Kullman of the League of Nations' Education Information Centre in Geneva. He provided some background to the educational work of the League, emphasizing "the 'facts' of world interdependence making world co-operation and world collective action not merely an ethical postulate but a tragic necessity of law," and stressed the role of education in sustaining the vision of the League.[43] Dr. Pierre Bovet, director of the Institut J. J. Rousseau outlined the threats and demands of nationalism in the modern world. Other speakers described the challenges posed by nationalism in the South African context.

One of the most prominent speakers was Fred Clarke, by this time head of the overseas division at the Institute of Education, London University, a post funded by the Carnegie Corporation.[44] His great impact at the conference seems to be explained by his ability to link local issues to the broader international context of the early 1930s and, in doing so, to open the way for future analysis of education. His key themes were: (a) the threat to democracy and education as presented by the rise of totalitarianism in Europe; (b) the relationship between the Commonwealth colonies of settlement and Europe, and (c) the colonizer/colonized issue in Africa, with specific reference to education. He defined issues of culture and transformation in "New Countries" for the most part in terms of dominion contexts.[45] He argued that there was a need to maintain a balance between preserving European tradition and formulating an innovative approach to the new educational environment.

It was left to the social anthropologists to make some of the most important political interventions in the debates. Many of those present were to make internationally significant contributions in their fields. These included Dr. Bronislaw Malinowski (head of the International Institute of African Languages and Culture, London School of Economics), Dr. Isaac Schapera (University of Cape Town), Dr. Monica Hunter (later Wilson) (International Insititute of African Languages and Cultures, London), Winifred and Alfred Hoernle (University of the Witwatersrand), and Professor W.W. Eiselen, Stellenbosch University.[46] The most important among these, from the point of view of the conference, was Malinowski. He delivered several addresses on the relation of education to problems of "culture contact" in Africa and emphasized the social role of education and culture in the colonial context.[47] He was at a significant stage of his career when he was questioning some of the established "rules" and conventions of structural functionalism,[48] acknowledging the political context of social change in colonial society, and increasingly arguing that "scientific study of African societies must be politically committed."[49] Just as the NEF had moved from a psychological/child-centered view of education in the 1920s to a more sociologically based view of education in the 1930s, so Malinowski moved from the structural functionalism of his early career[50] to a more politically oriented position between the 1930s and 1940s that recognized the social and economic significance of colonialism for interpreting the nature of African society in the contemporary context. Such a perspective necessarily had to acknowledge the repressive aspects of colonial rule, something both anthropologists and educationalists had been reluctant to countenance.

Engaging with the NEF debate from the outset, Malinowski argued that "education is bigger than schooling" and that it "is concerned not only with the development of the child's 'biological inheritance,' his mental endowment, but also with his cultural heritage and his place in society." He emphasized that while "the gap between the world inside the classroom and the world outside was great enough in Western society," it should be acknowledged that it is that much greater in African/colonial society. He pointed to the magnitude of the task of formal education in an African context and noted the lack of research and systematic thinking that had been characteristic of colonial education and the lack of expertise on how to graft formal or modern schooling to the traditions of the past.

This call for a "scientific approach" to education in Africa reflected in part the major theme of Loram and Malherbe and the management ethos of Teachers College, which had significantly influenced thinking about colonial education since the time of the first Phelps-Stokes reports in the early 1920s. Malinowski's thinking was in part in keeping with the adaptationist ideals of Jesse Jones and Loram when he "presented an indictment of the education offered by the colonial state and the missions." He argued that such schooling had "in the past been undertaken with easy assurance, on the assumption 'that what we feel is necessary and right for the African must be the best for the African,' and that it could be done through the school." While acknowledging that this schooling was often "dominated by the lofty and unselfish ideals of the Europeans," he argued that it was "nevertheless out of harmony with the real conditions," and helped to develop in school graduates a contempt for African culture, traditions, and society, often "causing a sense of inferiority and inadequacy."[51]

In keeping with the dominant eugenicist metaphor of the time, he emphasized the "disintegration" of African cultures in the context of contact with colonialism and the modern world and the challenges of "re-integration" of Africans through educational processes. In that context it was necessary to develop a system of education to meet the needs of the African child. He emphasized that "to educate a primitive community out of its culture—that which embodies and correlates tribal beliefs, ideas, values, organisation and pursuits—and to make it adopt integrally the culture of a different race, and of a much more highly differentiated society, is a gigantic task," which had been hugely underestimated to date.[52] Up to this point Malinowski appears to have shared ground with previous writers, but he then also introduced the reality of South Africa and other colonial states into the equation. He placed the political problem of colonial Africa clearly on the table.

> The African lives in a world which is politically subject, economically dependent, culturally spoon-fed, and molded by another race and another civilization. A considerable portion of his tribal lands has been alienated, the political independence of the whole society has been modified, and his traditional law, his economic pursuits, his religious ideas questioned.
>
> With this educational problem of the introduction of European education to the Africans there goes another: how is the child to secure the place for which this education fits him, in the face of the race prejudice, laws and attitudes connected with the color bar imposed by white communities to prevent social and economic equality between Black and White? It is unquestionably dangerous to expend all our generosity on giving (the African) a goodly measure of education, only to deprive him of the fruits thereof by force of law and political discrimination.
>
> To proceed in such a manner was, for Malinowski, to court tragedy.[53]

Although he favored a balance between European schooling and African education with "the necessity of cultural harmonizations," he argued that "all the evidence points to the conclusion that the African child responds well to the same type of schooling as the European." Malinowski returned to the conference theme with its idea of education as a "re-integrating agency" and a mechanism for reintegrating Africans into society on new terms that would enable them to manage their lives, economically and socially, in relation to the modern society that was emerging around them. At the same time it would enable them to regenerate African culture to cope with the needs of change and transition. There is insufficient space here to explore the contributions of the other anthropologists present, but Malinowski's broad focus provided a framework for many of their contributions as well. There is no record in the proceedings of the response to these views on the part of the sociologist Professor Hendrick Verwoerd, or the social anthropology professor W. W. M. Eiselen, both at the University of Stellenbosch, the ideologues of the future apartheid education policies, who also attended the conference. They might have taken courage from what they could have interpreted as support for their radical segregationism.[54]

With hindsight, the conference had a "modern" look. The section on African education considered topics and themes that were far removed from the individualistic, psychological, and Progressive "New Education" foci of the NEF conferences of the 1920s. The subagenda seems to have been a desire to display the ethos of modern education to a South African (and European) audience beginning to be faced with new forms of intolerance in the form of Nazism and Fascism. It must have been clear to many that these seeds of intolerance were also growing in the context of racially segregated South Africa. As Saul Dubow points out, the distinctions between segregationism in South Africa and the British colonial policy of Indirect Rule were blurred at this time, and the conference itself bore testimony to the difficulty of defining terms and practices associated with "culture contact" in terms that would set off the messages from the anthropologists present from the radical segregationist views represented by Holloway, Eiselen, and Verwoerd.

It all seemed to pave the way for the kinds of developments evident at Cheltenham two years later, though it was to be a long time before any international conference on education was again able to achieve this level of critical debate on educational policy in Africa.

Cheltenham Conference, 1936

The last of the great European interwar conferences of the NEF was held in Cheltenham, England, in 1936.[55] The theme, "Educational Foundations for Freedom in a Free Community," set the tone. This was a smaller conference, with the majority of delegates from the UK (461), France (72) and the United States (119). The attendance of no less than 77 members from the British Commonwealth was notable. There were no delegates from Germany, Austria, Russia, or Italy.[56] The individualistic and psychological tradition was further displaced by an increased focus on policy and administration. Fifty speakers were listed, a quarter of whom were from the academy, thus demonstrating that there was still significant participation for a wider audience of educationalists. Perhaps most significantly, the conference highlighted the links between the work of the NEF and the League of Nations.

Key speakers at the conference were R. H. Tawney, British Labour Party politician and advocate of *Secondary Education for All*;[57] Charles Freinet, a French communist; Pierre Bovet, director of the Institut J. J. Rousseau (Geneva); Michael Sadler, a prominent British educationalist (Oxford); and Percy Nunn, retired director of the Institute of Education (London University). The new president of the NEF, Fred Clarke, who had succeeded Nunn as director of the Institute of Education in 1936, was again a prominent influence. In drawing together the NEF and the Institute of Education, he was able to use the NEF to promote the institute and to highlight the social aspects of educational research.[58]

The Cheltenham NEF conference was a significant moment of interaction between the various networks of educationalists. Clarke arranged for a special focus on aspects of education in the colonial context in an apparent attempt to ensure a degree of continuity with the themes taken up at the Commonwealth Education Conference (1931) and the NEF conference in South Africa (1934). It is also clear that there was correspondence between J. H. Oldham and Betty Gibson at Edinburgh House and Dr. W. B. Mumford, representing the colonial department at the Institute of Education, over this issue.[59] In other words, the IMC network in the form of J. H. Oldham was to some extent present behind the scenes. This interaction between missionary education networks, philanthropic foundations, and the professional education forums of the NEF marks a significant development in the growth of an institutional culture relating to colonial or African education, and allowed those debates to enter the NEF forum alongside a greater commitment to the defense of democracy in Europe and Asia.

A special commission of the NEF on "African Thought on African Education" was a radical departure from the normal proceedings of the NEF conferences. It might have been an outcome of, or a reaction to, the Cape Town conference, where, according to his biographer, Clarke had made a considerable impact.[60]

The commission provided a space for a group of Africans to speak on the topic of "African Thought on African Education."[61] Reporting on the event, the journal *West Africa* commented: "At these meetings, the Europeans adopted an unfamiliar, more modest, attitude, that of listening while Africans delivered their own ideas of what was good for their countrymen."[62] The chairman of the commission was W. E. F. Ward, principal of Achimota College in the Gold Coast.[63] The "spokespersons" were not educational experts as such, but "educated Africans who had been studying in England during the past year." Each of them presented a paper. All contributions exhibited a forceful position on colonial education, broadly supporting a set of assumptions that are very familiar in the twenty-first century but were radical in the context of Cheltenham in 1936.[64]

Australasia, 1937

The 1937 NEF conference was held in Australasia in August and September, in association with the Australian Council for Educational Research. The hosting of the conference was proposed by K. S. Cunningham, an executive officer of the Australian Council for Educational Research, who had attended the South African conference three years earlier. It was held over a two-month period in a number of centers throughout New Zealand and Australia.[65] There were 21 speakers who gave all the 300 addresses.

Like the New Zealand event, the Australian "conference" took the form of a lecture tour by educational experts. Like the South African event, it provided wide scope for general participation and elicited great public attention and unprecedented interest in education, though, in contrast to the South African event, none of the discussions or comments are reported in the voluminous published proceedings.[66] Like many earlier NEF conferences, it was supported by the Carnegie Corporation of New York. In comparison to the South African and Cheltenham events it seems remarkable for its return to an earlier mindset.[67]

In contrast to the NEF of the 1920s, there was no strong international presence. Unlike the South African conference, there was no attempt at interdisciplinary interaction with social scientists from other fields, nor was there any attempt to engage with the issues that had been so prominent at the Commonwealth Conference and the South African conference regarding the non-dominion reaches of the Empire. Notably, there were no Africans, Indians, or Latin Americans present. The indigenous peoples of Australasia do not appear to have been mentioned.

In general, the conference, whatever its significance for Australian education, did little to advance the London, South African, and Cheltenham initiatives in broadening the scope of educational inquiry and research in relation to the wider international political challenges of education in the mid-1930s. In particular, the emergent debates on education in the rest of the British Commonwealth were neglected in the context of the determined initiative to reform the local education systems that were seen to be outmoded.

Perhaps of greatest significance for the themes explored here of interaction between the NEF networks and emergent networks and debates about education in the British Commonwealth was the visit to India by key delegates en route back to Europe. Bouvet, Hankin, Salter Davies, and Zilliacus broke their journey to make a three-month tour at a momentous time in Indian education. They attended the All-India Education Conference and met members of the Indian National Movement, including Gandhi, Nehru, and Tagore. Debates about basic education for the masses, the need to connect education to the environment of the child, and appeals for cultural and linguistic relevance for schooling provided the Europeans with new perspectives on their own continent and a wider perspective on international education issues.[68]

These contacts with Indian and African educators and issues since the 1931 British Commonwealth Education Conference, added to the increased attendance of Japanese and Chinese educators since the Nice conference, changed the tone of the deliberations. The growth of NEF branches in India and Japan was rapid in the mid-1930s, with large regional conferences being held by the local committees.[69]

The Parallel Development of Church-based, Philanthropic, and Pan-African Educational Networks in the Interwar Years

A central element in the shifting emphasis of the NEF conference deliberations is the emergence of a "scientific," research-based ethos and methodology on the social and political aspects of policy and education. Within the literature on these developments, there is little recognition of parallel networks that were both influenced by these changes and in turn impacted upon them in various ways. My particular emphasis here concerns unfolding debates about the nature of colonial and missionary education that were emerging in the parallel conferences of the IMC and other church-based networks in the United Kingdom, Europe, the United States, and elsewhere.

These discussions were in turn impacted upon by the increasing involvement of U.S. foundations in educational research and policy development in the period under review. Rockefeller, Spelman, Carnegie, Jeanes, Spelman, and Phelps-Stokes were among the leaders in the field and were instrumental in funding the government and independent research initiatives already mentioned. They also offered direct support to the missionary societies. At a remove from the field of policy development for the most part, but with considerable potential to influence the debates on colonial educational reform in the long term, was the pan-African movement, in the same period.

The emergence of a missionary education network in the late nineteenth century was of the greatest significance for the development of educational debate of all kinds, especially with regard to the colonial context. The whole missionary conference network was increasingly involved in the complex issues of how to link the propagation of Christianity to the economic, social, medical, and educational needs of colonized peoples, and how to work in the political framework of colonialism. These issues became the subject of a series of influential conferences, independent and in cooperation with the government, between 1910 and 1937.

J. H. Oldham, the IMC secretary in London, recognized the significance of defining the role of missionary educators more broadly if the missions were to retain their influence in a changing world. In 1923 he had written to the secretary of state for the colonies, W. G. A. Ormsby-Gore, chairman of ACNETA, that he was "anxious that the missionary societies should not work at cross purposes with the Government but cooperate to the utmost extent possible."[70] In his address to the IMC conference at Le Zoute in Belgium in 1926 on the theme of "The Christian Mission in Africa," he noted, "[W]ithin the lifetime and memory of those present, the opening up of the vast (African) interior had taken place. Now for the first time these peoples are being swept into the mainstream of human history, and what their development is to be under the impact of the new forces has become one of the main questions of the century."[71] By the 1930s, that question was to be as urgent for the Christian churches as dealing with the appalling consequences of the Depression or the rise of Fascism. As with the NEF, the implications were increasingly interpreted within the shared framework of ideas based on the guidelines set by the charter of the League of Nations.

There was a change in the nature of missionary work at this crossroads. Reformers such as Oldham recognized that the new path for missions was to engage in social development of various kinds as the best means to promote spiritual values, and that a working relationship with the colonial governments was the only means of gaining influence over the path of policy development. This was for him the key means for the missions to remain relevant in the field of education in a rapidly changing political context. Yet while there was clearly a need to work with colonial governments, others such as A. V. Murray (U.K.), Colin Leys (Kenya), and W. H. Macmillan (South Africa) saw this as increasingly problematic.[72] In the changing political climate in India, the West Indies, and British Africa, in particular, they argued that it was necessary for the missions to be more accommodating of the social needs of indigenous peoples if they were to retain an influence. In that political, social, and economic climate, the question of education or schooling often stood out as a key aspect of mission work. In the postwar era it was essential to link mission's policy to the evolution of the social doctrine of the League of Nations. The religious message of personal salvation came to be inextricably associated with the political message of national independence and freedom for many of the leaders of educational debate in the IMC network.

The great debates about educational reform in Africa were associated with the IMC's conferences at Edinburgh (1910), Le Zoute (1924), Jerusalem (1927), and Tambaram (1938), the first two Phelps-Stokes Commission reports on "Education in Africa" in the early 1920s, the Commonwealth Education conferences of 1911, 1923, 1927 and 1931, the NEF conferences in South Africa (1934) and Cheltenham (1936), and the great Oxford Life and Works conference on "Church, Community and State" of 1937.[73] Keith Clements notes that "the period from 1928 to 1934 was one of the most pivotal in the story of the ecumenical movement, of the European churches (particularly German) and of theology,"[74] and it produced a variety of influences in the field of education despite the fact that there seems to have been very little overlap between the attendance of individuals at the conferences from the various traditions.[75]

The two great conferences of 1936 and 1937 seemed to allow for the blending of the two areas of experience and research. Initiatives such as the International Institute of African Languages and Cultures (established in 1926 with the assistance of the Spelman Foundation) and the Overseas Division of London University's Institute of Education (established in 1934 and supported by a Carnegie Foundation grant of $67,500 for a three-year period) were concrete manifestations of the fertile blending of the various traditions aimed at providing a research background to educational policy development.

Whatever the reality of missionary activity on the ground, these conferences depict a highly nuanced debate. Far from being simple agents of imperialism, these missionary conferences provide considerable evidence of a desire to consult the best "scientific" advisers in relation to education. While the policies that emerged were far from perfect and often subject to harsh criticism by emergent African nationalists and white settlers, there can be little doubt that by the 1930s there were serious attempts to address issues of educational aims, provision, access, financing, quality, and curriculum through research and debate, and to move beyond the easy formulae of the Phelps-Stokes recommendations regarding adaptation. These networks were in part responsible for framing educational debate in terms that were fundamentally different from those that were emerging in South Africa to provide a foundation for apartheid education.

Le Zoute Conference, 1926

At the IMC conference at Le Zoute in 1926, aside from the wide variety of churchmen and women, including missionaries, there were no less than 36 "consultative members" invited specifically to provide various kinds of expert advice.[76] Among these were eminent researchers on Africa such as Professor R. L. Buell of Harvard University, Professor W. M. MacMillan of the University of the Witwatersrand, Professor J. Richter of Berlin University, Professor L. Verlaine of the Universite Coloniale d'Anvers, and Professor D. Westermann, director of the newly established International Institute of African Languages and Cultures in London. In addition, there were a number of officials of various kinds from the British, Belgian, and colonial governments in Africa. As within the NEF networks, there was sharp controversy and a search for scientific explanations and solutions to policy goals.

There were also medical experts and a range of missionary educationalists from the missionary fraternity, though few from a university/research background. The notable "consultative members" from the field of education were Thomas Jesse Jones, director of the Phelps-Stokes Fund of New York and editor of the two recent reports on education in Africa funded by the foundation; Charles Loram, head of the South African (government) Commission for Native Affairs and author of the influential book *Native Education in South Africa;* and Jackson Davis and E. G. Sage of the General Education Board, New York. The official representatives of the British government included Hans Vischer of ACNETA. There were also a number of medical and health experts and officials from the Carnegie, Laura Spelman, Rockefeller, Slater and Jeanes, and Phelps-Stokes foundations.

However, in terms of some of the themes explored above, there were very few Africans or African Americans present, and those who were most prominent were almost exclusively from South Africa and the United States. Consultative members in this category included Rev. Z. R. Mahabane, from the Orange Free State, the president of the African National Congress in South Africa; John L. Dube (uMafukunela), the principal of Ohlange Institute in Natal;[77] and Rev. N. T. Clerk, synod clerk of the Scottish Mission to the Gold Coast Colony. Other prominent black persons who were significant in educational debates were Max Yergan of the YMCA of the United States, at the time based at the University College of Fort Hare in the Cape Province and who, according to one source, "impressed the conference deeply by the sincerity and restraint of his contributions,"[78] and N. D. Oyerinde of the Nigerian Baptist Confederation.[79]

Of the 19 pages of conference recommendations, over 8 are taken up with educational issues. The topics covered included the Christian ideal in education, policy, curriculum, the education of women and girls, the medium of instruction, languages and literature, and religious education.[80] Although no direct references were made to the New Education or the NEF, the general direction of those educational principles was reflected in the Four Simples of T. J. Jones's writings and the Phelps-Stokes report, *Education in Africa,* which had strongly influenced mission education policy and the British Colonial Office policy document issued in 1924. As Kenneth King has shown, these views were pervasive at this time and were a reaction to the academic approach that had characterized Indian education during the nineteenth century. The conference decided that "the curriculum of all types of schools should be drawn up with complete awareness of the life of the community" and that "it is not the needs of a few individuals, nor the needs of this white man or that for clerks and artisans, that is to be regulative. The needs of the community are to be considered first and foremost (and were to) determine the curriculum and the conduct of the school."[81]

In all of this there was an attempt, overt or covert, conscious or unconscious, to conflate these ideas with key ideas and ideals of the New Education and Progressive Education. The aim of education was no longer to master the books and pass examinations "but the elevation of the tone and character of the community in which the school is placed."[82] Within this perspective the notion of *adapted education* was emphasized, based heavily on the version of vocational or community education for rural folk that had been developed in the postbellum South under the mentorship of Booker T. Washington.[83]

In 1929, A. Victor Murray, at that time a lecturer at Selly Oak Missionary Colleges in Birmingham, published the first edition of his monumental work on education in Africa, *The School in the Bush,* in which he attempted to locate the African educational debates of the time within a political and economic context. He managed to engage with the key aspects of the debate and to locate education within a wider context of political and sociological critique, and to engage directly with issues of Christian morality and duty in the field of education in a manner that eluded many of his contemporaries. Murray was extremely critical of *adapted* education for Africans, associated with the Phelps-Stokes reports and the Colonial Office report. He argued that the notion of adapted education for Africans, on the Southern model, presented a dangerous political trap for the educator. The argument that a special type of education was needed to accommodate the needs of rural communities in Africa was for Murray just another means of arguing for "education along his own lines," which could only too easily be a metaphor for "keeping him in his place." Murray was critical of the notion of community that was being used in this debate as he saw it as an abstraction that allowed "education to be disassociated from social structure."[84]

Oxford, 1937

The culmination of missionary discourse in the interwar era was centered on the great Life and Works ecumenical conference, with 425 delegates from over 40 countries, in Oxford in 1937 on "Church, Community and State."[85] The key question for the conference was, "What is the Christian message in the world of dictatorships and crumbling democracies?",[86] and the goal of the proceedings was to produce "a comprehensive and balanced statement of the present mind of the Church."[87] One of the seven reports presented at the conference was on the topic of "Church, Community and State in relation to Education."[88]

Although the meeting was essentially called to establish a renewed basis for Christian beliefs in a world threatened by war, it was also of some significance for educational thinking. There was little that had direct bearing on colonial education, the core of the deliberations established a basis for the revision of Protestant theology that was to survive the tribulations of World War II and become the basis for the World Council of Churches in 1960.

On the negative side, the conference, in its focus on Europe and the threat of war, entirely lost the emerging focus on education and society in the colonial world that had been so evident in South Africa and at Cheltenham. Unlike the 1926 conference of the IMC in Le Zoute, which had a whole category of invited delegates who were listed under the heading of "Consultative Members," the delegates to the Oxford conference included relatively few "experts" from outside the ecclesiastical world, and certainly none who would have qualified as experts on the "colour problem," the colonial world, or Africa, let alone in the field of education.

The only keynote speakers widely known outside ecclesiastical circles were Reinholt Niebuhr and Emile Brunner, European advocates of the "neo-orthodoxy" of Christian secularism, which called for Christians to engage ever more deeply in the "secular" sphere and "to risk themselves in manifesting God's purpose within the world,"[89] an issue with great potential relevance for the debate on the role of education. Other significant speakers were T. S. Eliot, and R. H. Tawney, both of whom attended the section on economics, which presented a strong challenge to the inequalities of life in modern capitalist society and a challenge to Christians to engage with these issues through organizations such as trade unions and community organizations. At the same time it needs to be noted that these issues were largely dealt with in a European or North American framework, and nothing appeared to have been said about the issues of colonialism and the "colour problem."

Few of the delegates who attended the Oxford conference seem to have been educationalists. The only name that is recognizable from the African missionary circuit is Rev. J. W. C. Dougall.[90] The only African delegate to attend these sessions was one of the three South Africans present, Rev. Solomon Mdala, a relatively unknown minister of the Wesleyan Methodist Church in Uitenhage.[91] Outside Europe and North America, there was also a small number of "delegates from other areas" (the official designation)—two each from China and Japan, and one each from the Dutch East Indies, Korea, and New Zealand.

In registering its position as a "Supra-national," "Supra-racial," and "Supra-class" fellowship, the conference committed the church to a defense of the individual in the face of cultural exclusion, nationalism, racism, and exploitation. The report on education stated that "where the community denies to some children an education which would enable them to develop their full power, or where it permits their exploitation in industry, the Church in God's name must enter the lists as their protector."[92] Although much of this was framed with an eye to the deepening political crisis in Europe and the Far East, factors influencing the present situation with regard to education were identified as the "secularization of modern life," the increasing power of nationalism, the strength of science and industry to give men faith in a secular outlook, and the social disintegration of traditional family life and culture in the context of urbanization and industrialization. All of this was seen to have relevance for education and curriculum as it implied a need to reexamine the task of the church in the field of education.

Although all of this has, in retrospect, a very European feel, it is clear that these debates, in particular those highlighting the politics of race and racism, had immense relevance to the colonial context in general and to colonial education in particular in the long run, even if these issues were not raised at the conference itself.[93] Although the Oxford conference of 1937 provided a grand finale to the saga of Christian education in the prewar era, it is disappointing as a milestone in colonial education as it failed to take advantage of the substantial gains that had been made with regard to the systematic analysis and research of the field of education in the colonial context since Le Zoute in 1926. Although the conference committed itself to the "development of new machinery for research and action," the actual proceedings showed little sign of such tendencies. As an index to this, Clements points out that the representation from the younger churches (including, presumably, churches in places such as Africa) was woefully small."[94] There seems to have been little of the excitement and sharp debate about the educational aspect of this conference by comparison with the 1931 and 1934 conferences, where a range of experts had been welcomed to engage in educational debate.

Conclusions

The events of the events outlined above raise a host of methodological questions about the nature of the history of education and about what counts as evidence in a field where much of the impact of the conferences and debates can only be inferred. There has been no attempt to measure the impact of the deliberations recorded here for the ongoing realities of policy development and implementation. Yet it is quite clear from the evidence presented and the debates outlined that these represent the materials that were to constitute educational discourse in the 1930s, a time when the foundations for modern educational research and methodology were being set. Central to the interpretation of these events is the question of how far these networks can be said to overlap with and complement each other. The links between metropolitan debates and an emergent ideology of education and democracy linked to the League of Nation seem very obvious. But the implications for the wider world of colonial Africa and India have not yet been adequately charted.

Some answers to the questions posed at the outset are embedded in these explorations. Are the debates about colonial education sui generis, and to be understood in terms of colonial exceptionalism via such notions as "cultural imperialism" or "racist colonial domination," or are they simply local variants of international policy developments? Are they part of a very specific kind of education designed for domination and colonialism or do they reflect the best international expertise in the area of education at the time as reflected by the NEF conference network? Are the emergent NEF guidelines for educational policy only relevant to the interpretation of policy developments in Europe, North America, and the Commonwealth Dominions, or are they a central feature for understanding colonial education in the 1930s? To put it the other way around, to what extent does an investigation of developments in colonial education help to throw light on aspects of educational reform in the imperial heartland?

These questions seem to lead to tentative answers about the emergent links between these debates. To a large extent, these links have been hidden in the past by the insularity of history of education research, which has often failed to make the connections required with other fields of inquiry. Clearly, the mainstream explorations of Progressive Education and the NEF have tended to dominate, but there has seldom been any appreciation of the extent to which changes in those debates were part of a wider interaction with other fields of inquiry, both within the field of education (missionary education or colonial education) and in the unfolding of interdisciplinary links with social anthropology, economics, and historical studies. In attempting to explore the history of colonial education in Southern Africa, this exercise has been part of a wider exercise to define the forces that provided the context of educational discourse and policy development in an era when fundamental changes were wrought in colonial education. These forces included both those leading toward the goals and objectives of mass education systems inspired by the ideals of democracy and the League of Nations, and by tendencies that were more closely linked to the racist and totalitarian objectives of the axis powers. This chapter has been part of an attempt to understand the complexities of that historical background.

Notes

1. This paper was first presented at the Southern African Comparative and History of Education Society (SACHES) meeting in Dar es Salaam, Tanzania, in September 2005.
2. Rita Hofstetter, "The Institute of Educational Sciences in Geneva, 1912–1948," *Paedagogica Historica* 40, nos. 5 and 6 (October 2004): 657–684.
3. Richard Aldrich, "The Training of Teachers and Educational Studies: The London Day Training College, 1902—1932," *Paedagogica Historica* 40, nos 5 and 6 (2004): 617–631; Aldrich, *The Institute of Education: 1902–2002: A Centenary History* (London: Institute of Education, 2002).
4. E. Fuchs, "Educational Sciences, Morality and Politics: International Educational Congresses in the Early Twentieth Century," *Paedagogica Historica* 40, nos. 5 and 6 (2004): 774; Percy Nunn, *Education: Its Data and First Principles* (London: E. Arnold, 1920); Bertrand Russell, *On Education, Especially in Early Childhood* (London: George Allen and Unwin, 1926) and *Education and the Modern World* (London: W. W. Norton, 1932); Alfred North Whitehead, *The Aims of Education and Other Essays* (New York: Macmillan, 1929); Fred Clarke, *Education and Social Change* (London: Sheldon, 1940).

5. E. G. Malherbe, ed., *Educational Adaptations in a Changing Society* (Cape Town: Juta, 1937).
6. Fuchs, "Educational Sciences, Morality and Politics," 766.
7. E. G. Malherbe, *Education in South Africa 1652–1922: A Critical Survey of the Development of Educational Administration in the Cape, Natal, Transvaal and the Orange Free State.* This was submitted for the PhD at Columbia University in 1924 and was published by Juta, Cape Town, in the following year.
8. The forerunner of the present Human Sciences Research Council (HSRC) of South Africa.
9. The *Fusion* government was a coalition of the South African Party and the National Party that had come into being in 1933 to cope with the effects of the Great Depression. For details on J. C. Smuts, see W. K. Hancock, *Smuts: The Fields of Force: 1919–1950* (New York: Cambridge University Press, 1968); for J. H. Hofmeyr, see Alan Paton, *South African Tragedy: The Life and Times of Jan Hofmeyr* (New York: Scribner, 1965).
10. World Fairs were held in 1867, 1878, 1889, 1893, 1900, 1904, and 1910. See Fuchs, "Educational Sciences, Morality and Politics," 759–784.
11. Kevin J. Brehony, "A New Education for a New Era: The Contribution of the Conferences of the New Education Fellowship to the Disciplinary Field of Education 1921–1938," *Paedagogica Historica* 40, nos. 5 and 6 (October 2004): 733–756; W. A. C. Stewart, *Progressive and Radicals in English Education: 1750–1970* (London: Macmillan, 1972): 364.
12. W. Rawson, *A New World in the Making: An International Survey of the New Education* (London: NEF, 1933), vi.
13. Stewart, *Progressives and Radicals,* 356–357; Brehony, "New Education," 742.
14. Cited by Brehony, "New Education," 742.
15. Stewart, *Progressives and Radicals,* 354.
16. Ibid., 376.
17. Brehony, "New Education," 751–754.
18. Stewart, *Progressives and Radicals,* 359.
19. W. Rawson, ed., *Education in a Changing Commonwealth* (London: NEF, 1931). Nunn was the president of the English Section of the NEF at this time.
20. W. Boyd and W. Rawson, *The Story of the New Education* (London: Heinemann, 1965), 89–90.
21. The Japanese invasion of Manchuria was to take place in September.
22. See P. Spear, *A History of India: 2,* chaps 15–17 (Harmondsworth: Penguin, 1975).
23. Rawson, *Changing Commionwealth,* iii.
24. The conference was addressed by Beatrice Ensor, the chairperson of the NEF, and the conference papers were edited by Wyatt Rawson, the assistant director. See Rawson, *Changing Commonwealth.*
25. Rawson, viii; E. G. Malherbe, "The New Education in a Changing Empire," in Rawson, *Changing Commonwealth,* 33–38. Reprinted as a pamphlet by van Schaik, Pretoria. Also see E. G. Malherbe, "Native Education in the Union of South Africa," *The Yearbook of Education 1933,* ed., E. Percy (London: Evans Brothers, 1933).
26. Rawson, *Changing Commonwealth,* 38–39.
27. Ibid., 39; A. V. Murray, *The School in the Bush* (Oxford: Longmans, Green, 1929); British Government, Cmd. 2374–1925.
28. Rawson, *Changing Commonwealth,* 26.

29. Judith Brown, *Modern India: The Origins of an Asian Democracy* (Delhi: Oxford University Press, 1985), 256.

30. Rawson, *Changing Commonwealth,* 53–58.

31. J. H. Oldham and B. D. Gibson, *The Remaking of Man in Africa* (Oxford: Oxford University Press, 1931).

32. IMC/ CBMS archive, Box 218 various for period 1929–1933.

33. K. Clements, *Faith on the Frontier: A Life of J H Oldham* (Edinburgh and Geneva: T & T Clark and WCC Publications, 1999), 249.

34. Rawson, *Changing Commonwealth,* 76.

35. Native Economic Commission report (Holloway Report). (U G–32: 1932). See also Cynthia Kros, "W. W. M. Eiselen: Architect of Bantu Education," in *Education under Apartheid,* ed. P. Kallaway (New York: Peter Lang, 2002), 39–52.

36. Richard Glotzer, "The Career of Mabel Carney: The Study of Race and Rural Development in the United States and South Africa," *Safundi* 10 (April 2003): 14–15.

37. E. G. Malherbe, *Report of the Carnegie Commission on the Poor White Problem in South Africa,* 5 vols. (Stellenbosch: Carnegie, 1932).

38. The Carnegie grant accounted for only a small proportion of the donations and grants to the conference, amounting to £160 out of a total of £1500. (Malherbe, *Educational Adaptations,* 540. For the context of such U. S. foundation grants in South Africa see Brahm Fleisch, "American Influences on the Development of Social and Educational Research in South Africa,1929–1934" (paper presented at the annual meeting of the History of Education Association, Atlanta, Georgia, 1992) and Richard Glotzer, "American Educational Research in the Dominions," *Educational Change,* 1997, 52–65. Glotzer notes that the Carnegie Corporation allocated a total of $500,000 to South African projects in 1929, the most significant portion of which was to fund the foundation of the SAIRR that was to play a leading role over the years in opposing racist legislation and political development (Glotzer, "Career of Mabel Carney," 12).

39. Malherbe, *Educational Adaptations,* 403–520.

40. J. Rheinallt Jones was a lecturer in native law and administration at the University of the Witwatersrand, editor of the journal *Bantu Studies,* and was to be a prominent member of the SAIRR, a major research initiative to promote civil rights for Africans funded by the Carnegie Corporation of New York and the Phelps-Stokes Fund. S.Dubow, *Racial Segregation and the Origins of Apartheid in South Africa, 1919–1936* (Oxford: Macmillan, 1989), 156; P. Rich, *White Power and the Liberal Conscience: Racial Segregation and South African Liberalism* (Johannesburg: Ravan, 1984).

41. Malherbe, *Educational Adaptations,* 404.

42. Glotzer, "Career of Mabel Carney," 20.

43. Malherbe, *Educational Adaptations,* 55.

44. Brehony, "New Education," 735; Aldrich," Training of Teachers," 617–631.

45. There do not seem to have been any delegates from Asia, Latin America, or the Pacific.

46. For a background to the field of social anthropology see A. Kuper, *Anthropologists and Anthropology: The British School 1922–1972* (New York: Pica Press, 1972) and Dubow, *Racial Segregation.*

47. Malinowski's two public addresses at the conference were subsequently revised and published under the title of "Native Education and Culture Contact" in the *International Review of Missions,* 25 (1936): 480–515. Despite the significance of his contribution on this topic, the Malinowski Papers show little evidence that he paid much systematic attention to the topic later in his career. (With thanks to Sue Donnelly, LSE Archives: Malinowski Papers).

48. *Structural functionalism* is an approach to anthropology that analyzes and describes the structure of societies as distinguished from their historical or comparative aspects.
49. Wendy James, "The Anthropologist as Reluctant Imperialist," *Anthropology and the Colonial Encounter,* ed. Talad Asad (London: Ithaca, 1973), 50–60.
50. B. Malinowski, *A Scientific Theory of Culture and Other Essays* (Chapel Hill: University of North Carolina Press, 1944/1961), 125–131.
51. Malherbe, *Educational Adaptations,* 423.
52. Ibid., 404.
53. Ibid., 424.
54. For more on the debates about race and intelligence/mental testing at the NEF conference, see Saul Dubow, *Illicit Union: Scientific Racism in Modern South Africa* (Johannesburg: Witwatersrand University, 1995), 215–223.
55. Conference papers: Wyatt Rawson, ed., *The Freedom We Seek: A Survey of the SocialImplications of the New Education* (London, NEF, 1937); Brehony, "New Education," 751.
56. Boyd and Rawson, *New Education,* 105–106.
57. Robins Davis, *The Grammar School* (Harmondsworth, Penguin, 1967), 61–62; J. S. Kaminsky, *A New History of Educational Philosophy* (Westport, CT.: Greenwood, 1993), 153–155.
58. Aldrich, "Training of Teachers," 630.
59. CBMS/IMC archive, Box 218, p.15, file New Education Fellowship 1936: Correspondence Betty D. Gibson/Dr W. B. Mumford, University of London, Institute of Education re biennial World Conference of the NEF, Cheltenham, August 1936. The detailed report of the conference that appeared in the journal *West Africa,* August 22 and 29, 1936, is also filed in the J. H. Oldham Papers, 10/2 39.
60. F. W. Mitchell, *Sir Fred Clarke: Master Teacher 1880–1952* (London: Longman, 1967), 65–66.
61. New Education Fellowship, Cheltenham Conference, Special Commission on "African Thought on African Education," see NEF and Colonial Department of the Institute of Education, London University, August, 10–13, 1936 and J. H. Oldham Papers 10/2 39; Special Commission of the NEF on "African Thought on African Education," *West Africa,* August, 22, 1936; Cheltenham Conference—NEF see CBMS/IMC Box 218, p. 15.
62. *West Africa,* August 22, 1936: 1155–1156; August 29, 1936: 1191–1194.
63. For details of Ward, see C. Whitehead, "W E F Ward," in *Colonial Educators: The British and Colonial Indian Education Service, 1858–1983* (London: Taurus, 2003).
64. The African delegates, and the papers presented at the conference, are listed in *West Africa,* August 22, 1936.
65. Brehony, "New Education," 751.
66. The conference was organized by the NEF and the Australian Council for Educational Research. The proceedings were edited by K. S. Cunningham under the title of *Education for Complete Living: The Challenge of Today; Proceedings of the NEF Conference Held in Australia from August 1 to September 20, 1937* (Melbourne: Melbourne University Press/OUP, 1938).
67. K. S. Cunningham, *Education for Complete Living,* xxix.
68. Boyd and Rawson, *New Education,* 102, 111.
69. Ibid., 101–103.
70. IMC/CBMS archive, Box 219, folder Education: Africa, Educational Policy Approach to the Government 1923 (in Memo to the CO).

71. Clements, *Faith on the Frontier,* 236; E. Smith, *The Christian Mission in Africa* (London, IMC, 1926).

72. Murray, *School in the Bush;* N. Leys, *Last Chance in Kenya* (London: Hogarth Press, 1931); W. H. Macmillan, *Africa Emergent: A Survey of Social, Political, and Economic Trends in British Africa* (London, Pelican, 1938).

73. J. H. Oldham *et al. The Churches Survey Their Task: Report on the Conference at Oxford, July 1937, on Church, Community and State* (London: George Allen & Unwin, 1937).

74. Clements, *Faith on the Frontier,* 267.

75. A comparison of the list of delegates from the NEF conferences, the Commonwealth Education conferences, and the IMC conferences reveals very few overlaps of attendance.

76. The papers of the Le Zoute conference were published in Smith, *The Christian Mission in Africa.*

77. These two South Africans were noted as significant consultative members who merited financial support from the IMC. See CBMS/IMC Box 217 (on Le Zoute).

78. Smith, *The Christian Mission in Africa,* 26.

79. The prominent South African journalist and politician Don Jabavu was not present but contributed a significant article to the special edition of the *International Review of Missions* dedicated to the conference.

80. Smith *The Christian Mission in Africa,* 109–119.

81. Ibid., 62.

82. Ibid., 63.

83. K. J. King, *Pan-Africanism and Education: A Study of Race, Philanthropy, and Education in the Southern States of America and East Africa* (Oxford, Clarendon, 1974).

84. Murray, *School in the Bush,* 306. See also the work of B Malinowski cited above.

85. J. H. Oldham, ed., *The Churches Survey Their Task: The Report of the Conference at Oxford, July 1937, on Church, Community and State,* vol. 8 (London: IMC, 1937).

86. Clements, *Faith on the Frontier,* 328.

87. Ibid., 330.

88. The other reports were on "Church and Community"; "Church and State"; "Church, Community and State in relation to the Economic Order"; "the Universal Church and the World of Nations." See Oldham, *Churches Survey Their Task,* 130–166.

89. Clements, *Faith on the Frontier,* 263.

90. He had been secretary to the Phelps-Stokes Commission in 1924 and member of the Colonial Office's Advisory Committee on African Education, head of the Jeanes School programme in Kenya, and was at this time an educational adviser to the Protestant missions of Kenya and Uganda.

91. He is listed in the *African Yearly Register* (ed. Mweli Skota). (Johannesburg: R.L. Esson, 1932).

92. Oldham, *Churches Survey Their Task,* 135.

93. There was a final prewar conference in this series held at Tambaram, India, in 1938. See *IMC, Tambaram/Madras Series, Madras, India, 1938; record of the IMC meeting December, 12–29, 1938,* 7 vols. (Oxford: IMC and Oxford University, 1939).

94. Clements, *Faith on the Frontier,* 328.

Part III

The Market, the State, and Transformations in Teaching

7

The Teaching Family, the State, and New Women in Nineteenth-Century South Australia

Kay Whitehead

In labor and social history, changes in the family as an economic and social unit have been the focus of considerable debate and research. Pat Hudson asserts that "most established accounts stress that industrialization in one way or another destroyed a family economy where work was centred on the home, carried out within domestic patriarchal social relations, where men and women made different yet equally indispensable contributions to household income and subsistence."[1] The replacement of the family economy with the male-breadwinner wage form occurred with great variation across time, the social spectrum, in different occupations, and in different parts of the world. For example, Wally Seccombe, Sonya Rose, and Louise Tilly and Joan Scott have focused on working-class families, while Leonore Davidoff, Catherine Hall, and Mary Ryan have made similar cases for middle-class households.[2] Although there are case studies of specific industries, particularly in Britain, there has been a recent call for further research on the male-breadwinner family in a wider range of occupations.[3] In the history of education the teaching family has also been the subject of research although it has not attracted the attention of labor and social historians.

Australian historians of education have clearly demonstrated that teaching was a family affair in the mid-nineteenth century, not only in terms of husband-and-wife teaching teams but also combinations of parents and children and women family members conducting private and state schools.[4] Marjorie Theobald has explored the interaction between teaching families and the burgeoning state, problematizing its relationship with married women in particular, and my research has considered the teaching family in the Lutheran and Catholic school systems as well as the state.[5] This chapter, however, focuses on state schoolteachers in nineteenth-century South Australia and explicates the position of single women teachers.

The theoretical framework for this chapter is based on Pavla Miller's comprehensive account of familial, economic, political, and educational change in Western countries over four centuries.[6] With Ian Davey she has theorized that there was a prolonged crisis of patriarchal as well as class relations during the transition from feudalism to industrial capitalism, a crisis eventually resolved by forging a new form of patriarchy based on the male-breadwinner wage form. Furthermore, Miller and Davey propose that there was a crisis in relations between children and adults within farming communities and proletarian families that prompted the traditional bastions of patriarchy—the churches and the state—to explore new forms of governance of children. They argue: "The origins of mass schooling, then, need to be located in the defensive experimentation by patriarchally structured churches [and subsequently the state] coping with a crisis in obedience originating in the gradual disintegration of patriarchalist social relations."[7]

While Miller and Davey have concentrated on students and their families, this chapter explicates patriarchal relations at the micro level by exploring the notion of the "teaching family," a social and economic unit shaped by the recruiting and governing practices of the South Australian state school system. Here, the teaching family encompasses husband-and-wife teaching teams, various combinations of parents and children, and all-female families.

This study charts the changing status of the teaching family before and after state intervention in schooling. First, it describes the structure of the teaching family prior to state intervention in schooling, identifying men's, women's, and children's social and economic contributions to the family unit. Then it shows that under the 1851 Education Act the teaching family was co-opted by the state to accommodate the demand for sex-segregated schooling. At the same time, men were privileged over women as teachers. With the introduction of compulsory schooling in 1875, the state began to employ teachers individually and differentiate their wages on the assumption that men would marry and that women would be single. In effect, the state, as employer, substituted the teaching family with married men and single women and marginalized married women. The main argument is that the reconstruction of teaching as waged labor shored up the patriarchal household by constructing men as sole breadwinners but also facilitated women teachers' economic and social independence. Furthermore, the case is made that the individuation of wages facilitated the economic and social conditions for single women teachers, who were discursively positioned as "new women," to unsettle patriarchal norms and contest the gender order.

The Teaching Family Economy

Family-based immigration underwrote the establishment of a British colony in South Australia in 1836. The 1834 South Australia Act stipulated that married male immigrants had to be accompanied by their spouses and children so that men would not be required to perform "the woman's part at home as well as the man's part in the field or workshop."[8] The family was conceptualized as an economic and social unit, and men's status as household heads and principal breadwinners was taken for granted. Women were assumed to be economically dependent, and not ladies of leisure. Rather, they would be engaged in productive activity within the vicinity of the home. Although women occupied subordinate social positions within the family and South Australian society generally, economic necessity made them "indispensable partners in the work of the household, farm and workshop as well as vital contributors to family incomes."[9]

In common with white women elsewhere, teaching was one of the key productive activities undertaken by women immigrants as part of the family economy. Housework, childcare, sewing, the provision of food, nursing the sick, and teaching were all nonmarket but tangible resources for the household.[10] Nora Young, for example, was responsible for her children's early literacy instruction and the ongoing education of her daughter. Although the Youngs were relatively well off, they could not afford a governess, and so Nora's teaching contributed economically and also culturally to the family unit.[11] Indeed, in many households, women taught the manners and morals that contributed to their family's status and respectability. Children and often other members of the extended family such as cousins were also integral to the family economy. In large households such as the Giles', women's work was shared among those who were old enough to assist as members of the family labor team. Myra, the eldest daughter, was responsible for the bulk of the sewing, Jane, the youngest daughter, was "chief baker" and their mother's main task was to teach the younger children. When their servant Patty's fiancé was killed, they reduced her workload and also taught her to read and write, thereby providing her with a valued skill that she would later use to generate income.[12] In all of these situations women's teaching was an economic and cultural contribution to the family, but rarely was it enumerated in census statistics, which focused mainly on men's paid work.

While teaching was assumed to be predominantly women's work within the household, both men and women across the social spectrum taught to generate income. For example, in the very early days of white settlement, Mr. Boots, an impoverished Methodist preacher, and Mrs. Boots taught working-class boys and girls separately in a day school, and the Giles' servant, Patty, used her literacy skills to earn her living as a teacher for a short time. There were day and boarding schools for middle-class boys conducted by men and a range of young ladies' schools, mostly in the hands of middle-class women.[13] While teaching was a temporary income-generating activity for some families such as Mr. and Mrs. Boots, for others, such as the Hilliers, it sustained the family for upward of twenty years. John and Jane Hillier, their four young children, and their servant arrived in South Australia in October 1837. John was an agriculturist. Within three months of their arrival, Jane signaled her intention to contribute to the family economy as a teacher by advertising that she had opened "a SCHOOL for a select and limited number of YOUNG LADIES."[14] In accepting other girls to educate with her daughters, Jane was simultaneously generating an income and contributing to her family's cultural capital. With John's death in 1843 she became the Hillier family's principal breadwinner.

By the mid-1840s Jane's school was one of the premier schools for young ladies in the colony, and her daughters had joined her as teachers. As an all-female teaching family the Hilliers offered "the usual English Education" with the accomplishments of French, music, and dancing. In so doing they were tapping into the demand among middle-class parents for their daughters to be educated as ladies.[15] Marjorie Theobald argues that the widespread concern for the moral welfare of girls in colonial society also underpinned the demand for them to be taught by women. Many middle-class girls received much of their education at home and in young ladies' schools under the tuition of women, and working-class parents were similarly reluctant to entrust their daughters to male teachers unless there was also a woman teacher in the school. Indeed, women were seen to be the moral guardians of girls.[16] Although many men also earned a living as teachers in mid-nineteenth-century South Australia, women like Jane Hillier had bargaining power *vis-à-vis* men in the education marketplace by virtue of their moral guardianship, and this would remain so throughout the century.

In essence, in mid-nineteenth-century South Australia, the family was an economic and social unit, and teaching was a productive activity by which men and women contributed to the family economy. As household heads and principal breadwinners, men mainly conducted mixed schools for working-class children and middle-class boys. Teaching, paid and unpaid, was also women's work. Women imparted the first lessons in literacy and the manners and morals that contributed to cultural capital. They passed on their expertise in the accomplishments in young ladies' schools and were positioned as moral guardians too. By 1850 the numbers of men and women temporarily or permanently contributing to their family economy as teachers prompted the Anglican bishop to remark: "There has been a perfect rush of Teachers of all sorts to the Colony. They and surveyors are as plenty as blackberries."[17]

"A Married Man Preferred—The Wife to Teach the Girls"

Notwithstanding this proliferation of private schools and teachers, there was a broad consensus among colonial leaders that the state should also intervene in the education of working-class children. However, the first two attempts to do so failed, and in 1851 the newly elected Legislative Council appointed a Select Committee to inquire into education and formulate new legislation. Under the 1851 Education Act the state was not empowered to establish its own schools. Instead, local communities were expected to establish schools, secure teachers who would provide nonsectarian instruction, and then apply for financial support from the state. If the applications were successful, teachers were granted a license and a stipend of £40 to supplement their tuition fees, and they were required to employ and pay their own staff. The Central Board of Education (hereafter called the Board) was convened to assess applications for licenses and supervise the operation of the 1851 Act, and Chief Inspector Wyatt was granted considerable discretionary powers in these matters.[18]

In Inspector Wyatt's first report and the discussions that took place at the Select Committee and on the Board, the productive activity of teaching was constructed as a profession, the aim being to attract people "of suitable character" into the new state school system.[19] An important aspect of professionalization was the establishment of a "normal school" to educate "persons of both sexes in the qualifications, intellectual and moral, necessary to make good and efficient teachers." The consensus was that the normal school should accept equal numbers of men and women. It was pointed out that women should acquire the same certificates as men and that they would teach literacy skills in infant schools. The discussion surrounding this proposal was not whether or not young women should be accepted but whether the sexes should be separated in the normal school, and it was agreed that the admixture of the sexes, especially between the ages of twelve and sixteen, was unwise. Contemporary anxieties about girls' moral fragility were reflected in discussions here. It was this discourse of moral danger that underpinned the decisions to make the normal school, and indeed the whole system, sex-segregated as far as possible. This discourse enabled middle-class women teachers, as moral guardians of girls and teachers of literacy, to be central participants in the development of the state school system, albeit with a marginal status.[20]

Given that one of the basic tenets of the 1851 Act was to make schools and nonsectarian instruction accessible to all children regardless of location, the Board opted to license a few relatively large single-sex schools in the capital city of Adelaide, and in less-populous rural areas where one mixed school was required, men were preferred as licensed teachers. Where there were sufficient students to establish single-sex schools, the Board granted separate licenses to men and women, some of whom were husbands and wives, thereby endorsing married women as contributors of paid labor to the family economy.[21] The Act did not provide specifically for differences in teachers' remuneration, and male teachers' obligations as principal breadwinners were implied in some stinging criticism about stipends. The £40 stipend was criticized as being a "woman's wage," and the critic went on to ask, "Why should our Government pay for woman's services at the same rate as an educated and efficient male teacher, who has more to do with his money, if he be a man at all?"[22] Although the Board had no power to amend stipends, it soon introduced measures that institutionalized male privilege and increased men's incomes *vis-à-vis* women teachers. For example, men earned bonuses for efficient teaching while women's stipends were kept to the minimum.[23]

Although much of the discussion about state schoolteachers was conducted with urban situations in mind, Inspector Wyatt realized that the vast majority of state schools would be small, mixed schools in rural districts. For these schools his decided preference was for married men as licensed teachers:

I think their moral character should be entirely without stain, and that they should be married men, especially as schools in the country must consist of both sexes and the supervision would in some measure be entrusted to the female as well as the master.[24]

To these ends he tried to license men first of all in mixed schools where the combined stipend and tuition fees were sufficient to support a teaching family, be that husband and wife, father and daughter, or male and female kin. Thus, women licensed teachers were more likely to be relegated to small, remote communities. Where there were enough students to license both a male and female teacher, women teachers were prevented from retaining boys over seven years of age (that is once they had learned the first lessons in literacy) in their schools, thereby guaranteeing men teachers' incomes. In essence, where possible, Inspector Wyatt tried to co-opt the teaching family as an economic and social unit to accommodate the need for sex-segregated schooling. In so doing he envisaged men as household heads and principal breadwinners, and women as moral guardians and contributors of paid and unpaid labor to the family economy. His thinking here reflected and powerfully enforced the traditional dominant patterns of family organization in the colony. In effect, the mid-nineteenth century teaching labor force in state schools was being conceptualized and constructed around the prevailing ideas of gender difference.

Rural communities supported Inspector Wyatt's preference for a teaching family in mixed schools. When trustees advertised for teachers, they stated their requirements succinctly: "A Married Man preferred—the Wife to Teach the Girls."[25] And when men applied to school trustees for employment, they assured them of a woman's presence, mostly wives or daughters, in the school. For example, in 1863 Augustus Winter wrote: "Should my application prove successful my exertions as a teacher would be supplemented by my wife and daughter; they have assisted me in the feminine department during the last fourteen years."[26] Applications for building assistance also stressed that boys and girls would be taught separately with the usual proposal being that the master for the boys would use the new building, leaving the old one for the mistress and the girls.[27] Furthermore, if there was no guarantee of a woman's presence as moral guardian and teacher of the girls in the school, the licensed teacher's livelihood was threatened. When George Needham's wife deserted him, the parents insisted that he employ a female assistant or they would withdraw their daughters.[28] In all of these cases women's labor as teachers did not appear in the statistics, for the Board only recorded and published the names of licensed teachers. However, women's labor and moral guardianship as both married and single members of teaching families was essential to the economic success of mixed schools, for their presence secured girls' attendance and their tuition fees.

In essence, both the Board and local communities (who, after all, actually appointed the teachers) wanted men and women as teachers for single-sex schools, where numbers made it feasible, and a teaching family for mixed schools. Most state schools were in the country, so men dominated the statistics as licensed teachers. In 1856, for example, women comprised only eighteen of the seventy-eight licensed teachers in country schools but outnumbered men slightly as licensed teachers in Adelaide.[29] Although these statistics seemed to show that teaching was predominantly men's work, women as productive members of teaching families were the hidden investment in state schools.

When an economic depression descended in the late 1860s, however, the Board was required to rationalize its provision of state schools. To this end it withdrew licenses from women teachers "who were conducting schools of an elementary character connected with schools for which male teachers were licensed, who in several cases were the husbands of licensed teachers." Amid storms of protest the Board withdrew licenses from teachers with fewer than forty students if the school was within two miles of a larger school. On the original list of fifty-six teachers, forty were women. This further entrenched men's privileged positions as principal breadwinners and left 222 men and 72 women licensed in the colony in 1870. This was the first and only time in the nineteenth century that women's marital status was formally cited as a justification for the removal of their license. Indeed, no marriage bar was included in the regulations governing the state school system until 1915. From 1870, however, no licenses were granted to the wives of licensed teachers, but many married women continued teaching in state schools as assistants while the 1851 Act was in operation.[30]

Sonya Rose states that nineteenth-century employers "patterned their workforces and hiring practices, structured work opportunities and managed their enterprises in ways that expressed pervasive meanings of gender difference, class relations and a developing ideology of family life."[31] Under the 1851 Act the educational state co-opted the teaching family as the ideal unit to accommodate the need for sex-segregated schooling. The state reinforced the patriarchal household by granting most licenses to men teachers as household heads and principal breadwinners in the family economy, thus protecting their positions. Women's presence in state schools was legitimated by their cultural capital as moral guardians of girls and as teachers of literacy, and by their need to contribute income and labor to the family economy. The withdrawal of licenses from the wives of licensed teachers, however, was the first signal that far-reaching change was afoot in the teaching workforce. By the early 1880s the teaching family had been dismantled, married women had all but disappeared from state schools, and all South Australian children between the ages of seven and thirteen were compelled to attend school.

Transforming the Teaching Family

In the early 1870s, agitation for the introduction of compulsory schooling gathered momentum. In this context the focus of legislators' attention turned to urban working class children, who, it was claimed, were not attending school. Amid these debates the Board was reconstituted, and John Anderson Hartley effectively replaced Inspector Wyatt. Hartley's influence was immediately apparent in the pressure on the government to build a large state school in the heart of Adelaide to cater to working-class children.

The head teachers and their assistants in the boys', girls,' and infants' departments at the new "Grote Street Model Schools," opened in 1874, were employed and paid by the Board. The salaries of the headmaster, Lewis Madley, headmistress, Lavinia Seabrooke, and infant mistress, Jane Stanes were £400,

£200, and £150, respectively, and male assistants were also paid more than females.[32] This was the first time the state in South Australia became an employer of teachers' waged labor, and the procedures established for this school set the precedent for subsequent legislation. The new salary scales are tangible evidence that gender was a significant consideration in the structuring of employment opportunities. Madley's salary assumed that he was a married household head with sole responsibility for supporting a family. There was no suggestion or implication that the state was securing the labor of his family as was the case with other state schoolteachers, or that he would have the prerogative to utilize family labor. The women's salaries indicate that the state assumed that they only needed enough income for immediate necessities such as food, clothing, and accommodation, that they were independent of family responsibilities, and that they would spend just a few years as waged workers prior to marriage.[33] Furthermore, at Grote Street, unrelated individuals were brought together in the workplace for the first time. They did not necessarily share common goals, and they were neither economically nor socially dependent on Lewis Madley. As Miller and Davey note, "The natural order of things with a male patriarch presiding over the labour of his family and other dependants, seemed to be turning on its head."[34]

Although all of these teachers had had considerable administrative experience, the Board upheld male privilege and informed Madley that he would control the entire school. Seabrooke protested strongly, and at a special meeting of the Board one week later, Madley's duties were altered. His authority over Seabrooke and Stanes was confirmed, but they were granted jurisdiction within their departments. Thus the headmaster was not totally empowered and the headmistress was not entirely powerless, but she had fewer resources to bargain with, and whatever power she exercised was on terms determined by men. Clearly, there was potential for increasing tension between men and women teachers as they pursued their careers.[35]

This new situation at Grote Street set the scene for countless battles inside late nineteenth-century state schools. From the outset there was fierce competition between the boys' and girls' departments as the headmaster and headmistresses attempted to consolidate their status publicly at the annual examinations, and in the other state schools the situation was equally fraught as the new Board moved swiftly to reorganize teachers' work.[36] Between January 1874 and the advent of compulsory education in December 1875, the occupational autonomy of the 217 men and 91 women who were licensed teachers ebbed quickly but not without many individual acts of resistance. The Board issued new instructions for the classification of pupils, teaching methodology, and timetables, and three new inspectors were appointed and given the authority to examine state-school students and report in detail to the Board.[37] In its reforming zeal the new Board eroded state schoolteachers' control of their daily labor and removed their prerogative to employ staff. Henceforth, licensed teachers were required to employ assistants and pupil teachers according to a formula, and all appointments had to be sanctioned by the Board. In effect, licensed teachers lost control of the use of family labor in schools.[38]

Late in 1875 both houses of parliament passed legislation for compulsory and secular state schooling. The Board was initially replaced by a Council of Education with Hartley as its president, but soon afterward, it too was disbanded, and Hartley became the Inspector-General, supported by a bevy of inspectors. Under the 1875 Education Act a highly centralized and bureaucratic Education Department was constructed with Inspector-General Hartley at the helm to control state schooling in South Australia.[39]

State schools with more than twenty students were designated as public schools, and they were to be conducted by trained teachers. The largest schools, most of which were in Adelaide, were divided into separate boys' and girls' departments, and the rules and salary differentials that had been implemented at Grote Street in 1874 were incorporated. However, most schools, especially in the country, were much smaller, so the most common staffing arrangement was that of a headmaster who was guaranteed a minimum salary of £150, and either a woman assistant whose salary was fixed at £40 or a pupil teacher. Men's privileged positions as head teachers were protected further by the regulation: "Should the average attendance be higher than 100, in any mixed school, the principal must be a master." Schools were required to have at least 100 students before an assistant was appointed. Given that almost all of the old licensed schools had fewer than 100 students, this basically denied women assistants, many of whom were the wives of licensed teachers, paid employment as public teachers in the new state school system.[40]

When the new regulations were published in January 1876, salaries were a major source of dissatisfaction. The complainants, mostly men, did not portray themselves as sole breadwinners. Rather they spoke as principal breadwinners of family economies where women's paid teaching labor was necessary for the family's survival. Several teachers, some of whom based their calculations on the aggregated family income, claimed that they would lose up to £100 under the new arrangements.[41] Given that they based their arguments on combined incomes, they had a vested interest in increasing the remuneration of women assistants and pupil teachers who, in many cases, were family members. Teachers were supported by the editors of the *Illustrated Adelaide News*, which claimed that "the labour or talent of women has been accounted as nothing," and by the *Register*, which argued that "the wives of school-masters are able to render material help in the schools, and are entitled to be paid according to the efficiency of their work."[42]

The regulations were amended in March and June 1876. Maintenance allowances for school cleaning were paid to all head teachers, thus increasing their total earnings, considerably in some cases. Statistics recorded at the end of 1876 indicated that the incomes of men and women head-teachers in public schools were commensurate with their former remuneration as licensed teachers.[43] Women assistants' salaries were increased slightly, and graduated salary scales were introduced for them. Without actually saying so, the Council also sanctioned the teaching labor of wives in public schools with a new regulation that stated: "In schools with an average attendance of 30 scholars and under 100 a Sewing Mistress may be allowed."[44] Although the numbers of sewing mistresses were recorded in the annual statistics, their appointments lapsed with the headmaster's removal from the school, and no service records of these women were kept. In effect, the state further institutionalized men as sole breadwinners by categorizing sewing mistresses, who were their wives in many cases, as nonteachers, and so blocking their permanent employment.

The lives of a number of teaching families changed dramatically under the new regulations. For example, Edward Catlow had been a licensed teacher in a small country school from 1865, and Augusta, his wife, was his assistant.[45] Under the new regulations he was employed as the head teacher with an annual salary of £231, but Augusta was only classified as a pupil teacher because the average attendance was seventy-five, that is, below the entitlement for a tenured assistant. After Edward's protests she was paid as a pupil teacher and sewing mistress until 1879 when their only daughter, Kate, was old enough to be employed as a monitor.[46] Augusta withdrew from income-generating labor, and Kate was inducted into the family's teaching enterprise, later to be classified as a pupil teacher.[47] Although Kate was able to replace her mother, the records show that the teaching family was rapidly dismantled in the state school system. Wives were first relegated to the temporary position of sewing mistress and then disappeared altogether.[48]

By the early 1880s married women were being refused employment as state schoolteachers. Although there was no marriage bar in the regulations, women teachers customarily resigned when they married, although not always of their own volition. The children of state schoolteachers who wanted to follow in their parents' footsteps found pupil-teacher apprenticeships in schools that were large enough to qualify for such positions, not necessarily their parents' schools. Indeed, Kate had to complete her pupil-teacher training in a large country school. Men teachers were paid salaries that assumed they were married and sole breadwinners, thereby confirming them as heads of patriarchal households, and the new Education Department's regulations shored up their privileged positions as male teachers *vis-à-vis* women teachers in the state school system. However, the aforementioned records also clearly show that they were no longer working with family members in their schools. Instead, the situation at Grote Street rapidly became the norm, and men were expected to work with single women with whom they had no familial relationship under increasingly invasive systems of inspection and accountability. Teachers such as Edward also lost status with the withdrawal of the first-class teacher's certificate that had been granted under the former legislation. These radical changes in the nature of employment and working conditions created such tensions that a government inquiry was initiated in 1881 and upgraded to a royal commission in 1882. Edward was one of the many male witnesses who spoke of the escalating tensions between men and women, as strangers were brought together in these new workplaces. However, Madley and other newly appointed headmasters defended Inspector-General Hartley. In the final report Hartley was exonerated, the reform agenda was confirmed, and within a few years the former licensed teachers had left the system—in Edward's case, through death.[49]

The process of transforming the teaching family had begun in 1870, when the wives of licensed teachers were deprived of their licenses, and it escalated under the 1875 Education Act when the state took control of employment and individuated wages. In effect the educational state was creating the sexual division of labor it took as normative by basing male teachers' salaries on the breadwinner wage and excluding married women. From 1875 teaching was constructed as the province of married men and single women, the tenure of the latter being dependent on their marital status. Kate Catlow represented the new generation of single-women teachers, that is, the first cohort of waged workers in the state school system. Her life as a state schoolteacher would be forged under working conditions very different from those during her parents' time.

Education " in the Hands of Unmarried Women"

Although women teachers were not a new phenomenon in state schools—indeed this chapter has shown that they had been the hidden investment in mid-nineteenth-century licensed schools—the new employment practices soon exposed their presence to public scrutiny. Under the old Board, only the names of licensed teachers, 70 percent of whom were men, were included in annual reports and in the minutes that were published in the newspapers. After 1875 these documents contained the names of individual teachers employed by the state, and women's numerical dominance was exposed. In January 1877 fifteen out of the seventeen candidates for the Training College entrance examination were women. In May 1877 a letter to the editor of the *Register* noted that men constituted only two of the nineteen appointments to state schools in the previous month. The correspondent continued: "If this goes on the arduous task of training our youth must fall wholly into the hands of women."[50] In 1879 one correspondent in the *Register* claimed that there were four or five female applicants for every male, bemoaned the absence of men, and predicted that "in no distant period the education of our youth will be almost entirely in the hands of unmarried women."[51] Inspectors also reported on the lack of male pupil-teachers and entrants to the Training College despite attempts to reserve places for them.[52] These concerns escalated to the extent that in 1885 Inspector-General Hartley addressed the issue in his annual report. In an appendix titled "Female Teachers," he marginalized dissenting voices, and to describe and justify their employment, he characterized women teachers as "nurturers" and teaching as a precursor to marriage and motherhood.[53] This strategy to quell opposition and endorse single women's waged employment worked. The lack of men teachers was not raised again in annual reports, although critics of the Education Department occasionally canvassed the issue.[54]

Contemporary commentators and, subsequently, historians, particularly in the United States, have termed this phenomenon the "feminization of teaching."[55] Most North American accounts portray teaching as men's work until the advent of state schooling, when men left teaching and the numbers and proportions of women increased to about seventy percent of the workforce by 1900. Feminization rates were higher in urban areas than in rural regions, and women were confined to the low-status, poorly paid junior grades while the remaining men occupied administrative positions. Concomitant with feminization was the focus on the nurturing woman teacher, fueled by influential commentators such as Catherine Beecher and Horace Mann. Rarely, however, do such accounts acknowledge that the first teachers of literacy were the large numbers of women who provided their unpaid labor as members of teaching families and conducted private schools before state schools were introduced, and who continued to do so throughout the nineteenth century.

What this chapter has shown is that in South Australia, teaching was women's work from the beginning of white settlement. Both married and single women continued their work, paid or unpaid, in a new context: the state school in the mid-nineteenth century. Teaching was not new work for women, but the state school was a new location for their labor. In addition, the feminization of teaching argument does not encapsulate the changes in the nature of men's or women's work, as the state individuated wages and fostered the sexual division of labor among teaching families. Connecting teaching with domesticity in the case of women further obscures the ideological and spatial separation of home and workplace that was integral to the reconstruction of the teaching workforce.

As far as men teachers were concerned, the state upheld their dominance in the patriarchal household by protecting their career paths, but the withdrawal of wives from visible paid work and also children, who were compelled to attend school, increased the pressure on men to be "good providers." As sole bread-winners, men teachers now had to earn sufficient income to support economically dependent wives and school-age children. They were no longer empowered to use their family's labor in their schools and had little say in the employment of the strangers with whom they worked. For example, John Peate complained bitterly about his woman assistant, blaming her for the school's poor aggregate result in the annual examination and his subsequent loss of salary: "I lost many children through Miss Jacob and my labour was thrown away." Although Inspector Stanton said that John's complaints were not entirely justified, he empathized that headmasters were "expected to bring their schools up to a certain standard with the aid of assistants in whose selection they had no voice."[56] While their wives were safely ensconced at home, men teachers had to work in close physical proximity to women who were not family members and over whom they could not exercise their traditional patriarchal authority as a father or husband. In fact, men teachers were confronted in the workplace by economically independent single women who did not share their family's goals. Leonore Davidoff argues:

> Given the structure of gender categories and their centrality to the nineteenth century concept of the family with its attendant male breadwinner, female housewife, non-working child roles, as well as the language of femininity and masculinity, it is not difficult to understand why women in public life posed such a threat to identity—for both men and women.[57]

In addition, women teachers did not necessarily see teaching as nurturing work. From John's perspective, Caroline was not an appropriately feminine teacher because her relations with children were not tender and motherly. Indeed, he claimed that she was "totally unfit for teaching in a public school. Her manner was harsh and she was so nagging with the children that they simply refused to do what she told them." It seems that at several levels Caroline contributed to John's anxieties as the sole breadwinner in a patriarchal household.

Women who taught in state schools were no longer part of family work teams but received separate earnings. Their wages were based on the assumption that they had no dependents and their conditions of employment assumed that they were free to transfer from school to school at the state's behest and to live and work apart from their families. Women teachers no longer worked under the authority of their father or husband in an extension of the family home but, in many cases, continued their work in purpose-built public institutions and negotiated their careers with unrelated men. The separation of home and workplace was both spatial and ideological in the case of Kate Catlow, who in 1883 left her parents' school to work among strangers at Mt. Gambier, the workplace where John and Caroline had clashed. She completed her pupil-teacher apprenticeship, graduated from the Training College, and was appointed to a country school in 1886. After one year she was transferred to Adelaide, where she spent brief periods in two large schools before being appointed to Norwood Model School in 1888. For the following ten years she worked with unrelated teachers in the girls' and infants' departments at Norwood, one of the largest state schools in Adelaide.[58] Wally Seccombe argues that in the case of women like Kate and Caroline, working away from home and earning their own wages "were beyond the bounds of patriarchal stricture. They conferred on women, if not yet in reality then at least potentially, a public presence and economic independence that flouted all traditional norms of women's place in the family households of their fathers and husbands."[59]

In addition, a substantial minority of women remained in the teaching workforce for many years, thereby retaining their independence well beyond societal expectations that they would marry by about the age of twenty-five. Kate did not resign to marry until she was thirty-two years old, and Caroline rejected housewifery in the patriarchal household completely, ultimately becoming the owner and headmistress of a prestigious private girls' school.[60]

Apprehensions were not linked only to women teachers' economic potential but also to the increase in their public profile as they participated in a range of social and political activities. Women now played sports such as tennis and rode bicycles for pleasure as well as a means of transport. Their leisure time was increasingly spent outside the family circle and thus was not subject to patriarchal control. Their increasing involvement in a range of political and social reform activities, for example, the suffrage campaign that culminated in South Australian women being enfranchised in 1894, also challenged male control of the public sphere. Women teachers, especially in Adelaide, spent their working days in the company of women, and their associational activity was both political and social. Kate's memberships in the Old Students' Association and teachers' union, for example, provided her with the opportunities to attend social events as well as meetings on professional matters and to forge friendships with like-minded women.[61] Then there was the issue of women teachers' domestic arrangements. Teachers' conditions of employment, namely, the requirement to transfer from school to school in pursuit of a career, prevented many women from living with their families but also provided new options for them to assert their independence and choose alternative living arrangements. In country schools women teachers sometimes lived alone but more often boarded with local families, while in Adelaide, boarding houses for women proliferated, some teachers lived alone, and some established households with other women.[62] For a small but significant number of women these domestic arrangements were permanent for they never married, choosing instead to make teaching their life's work. Most of the teachers who reached senior positions in state schools remained single. Education Department records indicate that there was a core of women teachers who taught for more than twenty years and that there were many whose careers exceeded ten years, among them Kate Catlow who resigned to marry after a twelve-year career. The evidence suggests that women teachers were not as transient as Inspector-General Hartley's rhetoric implied, and it could be that sizeable numbers were delaying or rejecting marriage as women's ultimate destiny.[63] As Katie Holmes notes:

> Singleness, accompanied by sufficient money, could offer women the opportunity to create a new lifestyle, a new identity: their vision of the single woman involved imagining another self, a self free from the physical, financial and emotional bondage of marriage. Single women could be agents of their own lives.[64]

In fact, the contradictory practices of the state are evident in the careers and private lives of women teachers in state schools. The state marginalized married women as teachers and portrayed teaching as a preparation for motherhood to justify single women's employment as waged workers. In so doing, it confirmed the patriarchal household and participated in the social construction of married women's dependency in late nineteenth-century society. Yet it also fostered the conditions for single women to be socially and economically independent. Many spent their careers in state schools, but some, like Caroline Jacob, used their credentials and experience to establish successful private schools. Whatever the context, waged employment conferred on women teachers sufficient economic independence to make marriage a choice rather than an economic necessity. It would seem that some women teachers were utilizing their economic independence to explore new options in their private lives.[65]

The appearance of the first generation of middle-class women who had been educated and employed as waged workers in the 1870s, and who rose to prominence in the 1880s, was the subject of public debate by the 1890s in Australia and elsewhere. The so-called woman question reflected the tensions surrounding women's visibility in the workplace as well as the decisions they were making about their private lives. The statistics indicated a significant increase in the numbers of single women in the paid workforce. Some found their niches in the professions, with a growing band of women employed in manufacturing and commerce. Although it was now acceptable for women to undertake paid work prior to marriage, the statistics revealed an increasing age at marriage, a declining birthrate, and an increasing proportion of women never marrying.[66] Furthermore, the numbers of single women were rising at "a time when there was a great concern about the declining birthrate, racial purity and a growing fascination with eugenics."[67] Of particular concern were the numbers of middle-class women who seemed disinclined to marry and reproduce, preferring instead to remain in paid work, live separately from their families, and participate in a host of other public activities.

By the late nineteenth century the term "new woman" had been coined to describe this cohort of well-educated, socially and economically independent single women. "The newspapers and magazines began to talk of the 'New Woman,' who was modern, capable and independent, informed of the affairs of the day, demanded political rights and education for women, and the right to earn her living."[68] Teachers, the largest and most visible group of women in professional employment, were identified as the vanguard of new women, and the anxieties surrounding their presence in state schools replicate those of the cohort generally. Although the presence of women as teachers in state schools was not a new phenomenon, their work became more visible with the advent of compulsory education. The numbers of single women teachers, now waged employees working in purpose-built institutions, gave rise to adverse publicity with the introduction of compulsory schooling. By sheer virtue of their numbers, single women teachers challenged the marginal status ascribed to them even though they were relegated to subordinate positions in state schools. Of even greater concern were the new social relations that attended the ideological and spatial separation of home and workplace. Working alongside unrelated married men and standing in stark contrast to male teachers' wives, women teachers challenged patriarchal governance in their day-to-day work and political and social activity. Then there were the significant numbers of women teachers who seemed not to imbibe the discourse of teaching as preparation for motherhood. Some teachers spent long periods as waged workers before marrying, while others never married, opting to spend their lives in the company of other women or family members and seemingly rejecting men, marriage, and maternity. Maternity was an especially significant threat because teachers, as well-educated middle-class women, were perceived to not only be rejecting patriarchy but also contributing to the demise of the white race. In effect, women teachers as new women were mounting a challenge to the gender order in their work and in their private lives in the late nineteenth century.[69]

Conclusion

By 1900 the state school system was the dominant provider of elementary education in South Australia, and most children between the ages of seven and thirteen had been brought under some form of school governance. The construction of the system had been a slow and uneven process, beginning in the 1850s and gathering momentum under the 1875 Education Act, with the state gaining the ascendancy in agricultural districts long before it did so in Adelaide, where there was a significant working-class population. These changes in the education landscape were achieved in the main with the labor of women teachers—the so-called feminization of teaching—and in South Australia as well as North America, single women constituted 70 percent of the waged workers in state schools by 1900.

However, by focusing on the reconstruction of the teaching workforce, this study has shown that women teachers had first colonized mid-nineteenth-century schools as paid and unpaid members of teaching families, their presence obscured in the statistics and by the recruiting and governing practices of the state school system. Under the 1851 and 1875 acts, the educational state upheld men's positions as household heads and as principal, and then sole, breadwinners, and institutionalized male privilege in state schools. Yet it was also involved in constructing the very gender relations it took as normative when it introduced teaching as waged labor. Privileging men as teachers and marginalizing married women shored up the patriarchal household, but it also facilitated the economic and social conditions for single-women teachers to construct their lives differently.

The expansion of state schooling in last quarter of the nineteenth century conferred on single women teachers opportunities to perform useful work in public institutions, gain modest economic independence, and make choices about the way they would construct their private lives. Notwithstanding the justification of their work as a preparation for domesticity, by 1900 significant numbers of women teachers opted to delay or reject marriage and actively pursue their careers. As this chapter has shown, their presence in schools and in society was not only conspicuous but also profoundly unsettling, so much so that as "new women," they were seen to be threatening the gender order by the end of the century.

Notes

1. Pat Hudson, "Women and Industrialization," in *Women's History: Britain, 1850–1945*, ed. June Purvis (London: UCL Press, 1996), 23.
2. Wally Seccombe, "Patriarchy Stabilized: The Construction of the Male Breadwinner Wage Form in Nineteenth-Century Britain," *Social History* 11, no. 1 (1986): 53–76; Wally Seccombe, *Weathering the Storm: Working Class Families From the Industrial Revolution to the Fertility Decline* (London: Verso, 1993); Louise Tilly and Joan Scott, *Women, Work and Family*, 2nd ed. (New York: Holt, Rinehart and Winston, 1987); Sonya Rose, *Limited Livelihoods: Gender and Class in Nineteenth Century England* (London: Routledge, 1992); Leonore Davidoff, *Worlds Between: Historical Perspectives on Gender and Class* (Cambridge: Polity Press, 1995); Mary Ryan, *Cradle of the Middle Class: The Family in Oneida County New York, 1790–1865* (New York: Cambridge University Press, 1985); Leonore Davidoff and Catherine Hall, *Family Fortunes: Men and Women of the English Middle Class, 1780–1850* (London: Hutchinson, 1987).

3. Angelique Janssens, "The Rise and Decline of the Male Breadwinner Family? An Overview of the Debate," in *The Rise and Decline of the Male Breadwinner Family?* ed. Angelique Janssens (Cambridge: Cambridge University Press, 1997), 2.

4. Marion Amies, "The Career of a Colonial Schoolmistress," in *Melbourne Studies in Education 1984,* ed. Imelda Palmer (Melbourne: Melbourne University Press, 1984); Richard Selleck, "A Goldfields Family," and Marjorie Theobald, "Women's Teaching Labour, the Family and the State in Nineteenth Century Victoria," in *Family, School and State in Australian History,* ed. Marjorie Theobald and Richard Selleck (Sydney: Allen and Unwin, 1990); Richard Selleck and Martin Sullivan, eds., *Not So Eminent Victorians* (Melbourne: Melbourne University Press, 1984).

5. Marjorie Theobald, *Knowing Women: Origins of Women's Education in Nineteenth Century Australia* (Melbourne: Cambridge University Press, 1996), chap. 6; Marjorie Theobald, "Writing the Lives of Women Teachers: Problems and Possibilities," in *Melbourne Studies in Education 1993,* ed. Lyn Yates (Glen Waverly: La Trobe University Press, 1993); Kay Whitehead, *The New Women Teachers Come Along: Transforming Teaching in the Nineteenth Century,* ANZHES Monograph Series, no. 2 (Sydney: Australian and New Zealand History of Education Society, 2003); Kay Whitehead, "'Religious First—Teachers Second': Catholic Elementary Schooling in Nineteenth-Century South Australia," *Change: Transformations in Education* 4, no. 1 (2001): 63–75; Kay Whitehead, "German Schools and Teachers in Nineteenth Century South Australia," *Paedagogica Historica* 37, no. 1 (2001): 55–68.

6. Pavla Miller, *Transformations of Patriarchy in the West, 1500–1900* (Bloomington: Indiana University Press, 1998).

7. Pavla Miller and Ian Davey, "Family Formation, Schooling and the Patriarchal State," in *Family, School and State in Australian History,* ed. Marjorie Theobald and Richard Selleck (Sydney: Allen and Unwin, 1990), 18.

8. Pavla Miller, *Long Division: State Schooling in South Australian Society* (Adelaide: Wakefield Press, 1986), 4; see also Helen Jones, *In Her Own Name: A History of Women in South Australia from 1836,* 2nd ed. (Adelaide: Wakefield Press, 1994), 49.

9. Pat Hudson and W. Robert Lee, "Women's Work and the Family Economy in Historical Perspective," in *Women's Work and the Family Economy in Historical Perspective,* ed. Pat Hudson and W. Robert Lee (Manchester: Manchester University Press, 1990), 3.

10. Harriet Bradley, *Men's Work, Women's Work: A Sociological History of the Sexual Division of Labour in Employment* (Cambridge: Polity Press, 1989), 8.

11. Nora Young to Mrs. Hext Bodger, 17 July 1870, Letters of Lady Charlotte Bacon, PRG 541, Mortlock Library of South Australiana (MLSA).

12. Jane Isabella Watts, *Family Life in South Australia: Fifty-Three Years Ago* (Adelaide: W. K. Thomas, 1890), 23–26.

13. Louise Brown, *A Book of South Australia: Women in the First Hundred Years* (Adelaide: Rigby, 1936), 37.

14. *South Australian Gazette and Colonial Register,* 20 January 1838, 1; Reg Butler and Alan Phillips, *Register Personal Notices* (Gumeracha: Gould Books, 1990), 220, 222.

15. *South Australian Almanack 1843* (Adelaide: Allen Publisher, 1843), 12; Marjorie Theobald, "Women and Schools in Colonial Victoria 1840–1910" (PhD thesis, Monash University, 1985), 14, 19, 200.

16. Marjorie Theobald, "Discourse of Danger: Gender and the History of Elementary Schooling in Australia, 1850–1880," *Historical Studies in Education* 1, no. 1 (1989): 29–52.

17. Augustus Short to Millicent Short, 24 May 1850, PRG 160/53, MLSA.

18. Malcolm Vick, "The Central Board of Education, 1852–1875" (MEd thesis, University of Adelaide, 1981), 3, 50–51, 131–132; Miller, *Long Division,* 24–25.

19. Report from the Select Committee of the Legislative Council appointed to consider the propriety of bringing in a General Education Measure, *South Australia: Votes and Proceedings* (hereafter *SAV&P*), Legislative Council, 1851, 1, 3, 5, 7–8; Vick, "The Central Board of Education," 108. For an expanded discussion, see Kay Whitehead, "From Youth to 'Greatest Pedagogue': William Cawthorne and the Construction of a Teaching Profession in Mid-Nineteenth Century South Australia," *History of Education* 28, no. 4 (1999): 395–412. For further discussion of discourses of professionalism, see Robert Gidney and Winnifred Millar, *Professional Gentlemen: The Professions in Nineteenth Century Ontario* (Toronto: University of Toronto Press, 1994), 5, 205–207; Burton Bledstein, *The Culture of Professionalism: The Middle Class and the Development of Higher Education in America* (New York: Norton and Co., 1976), 7, 80, 86, 111–112; Geraldine Clifford, "Man/Woman/Teacher: Gender, Family and Career in American Educational History," in *American Teachers: Histories of a Profession at Work,* ed. Donald Warren (New York: Macmillan, 1989), 321.

20. Report from the Select Committee, *SAV&P,* Legislative Council, 1851, 1, 3, 5, 7–8; *South Australian Government Gazette (SAGG),* 17 August 1851, 557.

21. Minutes, Central Board of Education, GRG 50/1, no. 30, 8671, State Records of South Australia (SRSA). For a similar situation see Geraldine Clifford, "Lady Teachers and Politics in the United States, 1850–1930," in *Teachers: The Culture and Politics of Work,* ed. Martin Lawn and Gerald Grace (London: Falmer Press, 1987), 6–7.

22. *Register,* 30 August 1853.

23. *SAGG,* 15 February 1855, 131.

24. Report from the Select Committee, *SAV&P,* Legislative Council, 1851, 17.

25. *Register,* 12 January 1871, 5 January 1867, 28 April 1852.

26. Augustus Winter to Trustees of Goolwa School, GRG 18/113/34, SRSA.

27. Henry Parsons to Central Board of Education, 20 July 1863, GRG 18/113/38, SRSA.

28. Minutes, Central Board of Education, GRG 50/1, nos. 5147, 8981, SRSA.

29. *Register,* 1 September 1856.

30. Minutes, Central Board of Education, GRG 50/1, no. 946, SRSA; see Reports of the Education Board, *South Australian Parliamentary Papers (SAPP)* 1869–70, no. 19, 8; SAPP 1870–71, no. 18, 8; *SAPP* 1871, no. 22, 1–4; Vick, "The Central Board of Education," 246.

31. Rose, *Limited Livelihoods,* 34.

32. Minutes, Central Board of Education, GRG 50/1, nos. 2408, 2426, 2447, 2448, 2540, 2556, 2571, SRSA.

33. For further discussion, see Alice Kessler-Harris, *A Woman's Wage: Historical Meanings and Social Consequences* (Lexington: University Press of Kentucky, 1990), chap. 1.

34. Miller and Davey, "Family Formation, Schooling and the Patriarchal State," 16; see also Sonya Rose, "Gender at Work: Sex, Class and Industrial Capitalism," *History Workshop Journal* 21 (Spring 1986): 113–131; Tilly and Scott, *Women, Work and Family,* 9.

35. Minutes, Central Board of Education, GRG 50/1, nos. 2609, 2626, 2632, SRSA; Report of the Education Board, *SAPP* 1874, nos. 24, 23. For a more general discussion, see Patricia Grimshaw, Marilyn Lake, Ann McGrath, and Marian Quartly, *Creating a Nation* (Melbourne: McPhee Gribble, 1994), xx.

36. *Register,* July 18 and July 20, 1874.

37. Report of the Education Board, *SAPP* 1874, no. 24, 3; *SAPP* 1875, no. 26, 3–13. For a discussion of inspectors' role in articulating the need for change, see Bruce Curtis, *True Government by Choice Men?: Inspection, Education and State Formation in Canada West* (Toronto: University of Toronto Press, 1992), 30–31.

38. Report of the Education Board, *SAPP* 1875, no. 26, 2.

39. Miller, *Long Division*, chap. 3.

40. Education Regulations, *SAPP* 1876, no. 21, 2, 4–7; Boucaut Papers on Education, 1876–77, V99, MLSA; *Register*, 14 February 1876.

41. *Register*, 3, 8, 11, 14, 17, 26, 29 January, 1, 11, 12, 14, 15, 18 February 1876; notes refer Education Council and New Regulations, Boucaut Papers, V132, MLSA; Boucaut Papers on Education, 1876–77, V99, MLSA; Council of Education minutes, GRG 50/1, 1876, nos. 111, 150, 252, 277, 308, SRSA; John Jones to Council of Education, December 1876, GRG 50/3/1876/2456, SRSA.

42. *Register*, 26 January 1876; see also *Illustrated Adelaide News*, 15 January 1876, *Register*, 14 January 1876.

43. Report of the Council of Education, *SAPP* 1877, no. 34, 1; *Register*, 8 April 1876, 17 July 1876; Incomes of School Teachers 1875–76, SAPP 1878, no. 42, 1–15.

44. Education Regulations, *SAPP* 1876, no. 21, 5.

45. Report of the Education Board, *SAPP* 1866–67, no. 41, p. 10; *Observer*, 4 April 1885; List of Members of the Adelaide Philosophical Society, 1864–75, SRG 10/14, MLSA; Butler and Phillips, *Register Personal Notices*, 149.

46. Edward Catlow to Council of Education, October 1876, GRG 50/3/1876/1893, SRSA; Incomes of School Teachers 1875–76, *SAPP* 1878, no. 42, 2, 7.

47. Minutes, Council of Education, GRG 50/1, nos 1958, 2070, 4516, 4962, SRSA; Alphabetical Register of Files on the Appointment of Teachers, 1876–87, GRG 18/116, SRSA; Tilly and Scott also note the persistence of these family strategies under the new economic order; see Tilly and Scott, *Women, Work and Family*, 232.

48. Compare, for example, the lists of state schoolteachers in 1875, 1876, and 1880, Incomes of School Teachers, 1875–76, *SAPP* 1878, no. 42, 1–15; Teachers of Public Schools, *SAPP* 1881, no. 90, 1–10.

49. Progress Report of the Select Committee of the House of Assembly on Education, 1881, *SAPP* 1881, no. 122, 72; Progress Report of the Commission on the Working of the Education Acts, 1882, *SAPP* 1882, no. 27, 1–2; Miller, *Long Division*, 38–39.

50. *Register*, 1 January 1877, 4 May 1877.

51. *Register*, 19 June 1879.

52. Council of Education Report, *SAPP* 1877, no. 34, 24; Council of Education Report, *SAPP* 1878, no. 40, 8, 32–33; Boucaut Papers on Education 1876–77, V99, MLSA; Mr. Madley's Report, *SAPP* 1884, no. 44, 4, 24; Mr. Madley's Report, *SAPP* 1884–85, no. 44, 25–26; Inspector Dewhirst's and Mr. Madley's Reports, *SAPP* 1886, no. 44, 1, 18.

53. Inspector-General's Report, *SAPP* 1886, no. 44, xx.

54. *Register*, 30 May 1888, 15 July 1889.

55. Jackie Blount, "Spinsters, Bachelors, and Other Gender Transgressors in School Employment, 1850–1990," *Review of Educational Research* 70, no. 1 (2000): 83–101; Laura Haniford, "Crisis in Masculinity: Changing Views of Women Teachers, 1870–1930" (paper presented at the annual conference of the American Educational Research Association, New Orleans, April 2002); Michael Apple, *Teachers and Texts: A Political Economy of Class and Gender Relations in Education* (New York: Routledge and Kegan Paul, 1986), 58–76; Alison Prentice and Marjorie Theobald, "The Historiography of Women Teachers: A Retrospect," in *Women Who Taught: Perspectives*

on the History of Women and Teaching, ed. Alison Prentice and Marjorie Theobald (Toronto: University of Toronto Press, 1991); Marta Danylewycz, Beth Light, and Alison Prentice, "The Evolution of the Sexual Division of Labor in Teaching: Nineteenth Century Ontario and Quebec Case Study," in *Women and Education,* 2nd ed., ed. Jane Gaskell and Arlene McLaren (Alberta: Detselig Enterprises Ltd., 1991), 33–36; Miller, *Long Division,* 52–53, 372; Barry Bergan, "Only a Schoolmaster: Gender, Class and the Effort to Professionalise Elementary Teaching in England 1870–1910," *History of Education Quarterly* 22, no. 1 (1982): 1–21. For a contrasting account see Noeline Kyle, "Woman's 'Natural Mission' but Man's Real Domain: The Masculinisation of the State Elementary Teaching Service in New South Wales," in *Battlers and Bluestockings: Women's Place in Australian Education,* ed. Sandra Taylor and Miriam Henry (Canberra: Australian College of Education, 1988).

56. Progress Report of the Commission on the Working of the Education Acts, 1882, *SAPP* 1882, no. 27, 122, 126, 148; Inspector Stanton's Report, *SAPP* 1881, no. 44, 14.

57. Davidoff, *Worlds Between,* 263; see also Haniford, "Crisis in Masculinity," 13–16.

58. For Kate's teaching record, see Teachers' Classification Board and Teachers' History Sheets 1882–1960, GRG 18/167, SRSA.

59. Seccombe, "Patriarchy Stabilized," 66.

60. Helen Jones, *Nothing Seemed Impossible: Women's Education and Social Change in South Australia from 1836* (St. Lucia: University of Queensland Press, 1985), 76–84.

61. *Education Gazette,* May 1885, 16; June 1886, 39; September 1896, 103. For general discussion, see Clifford, "Lady Teachers and Politics," 15.

62. This was also the case in America; see Blount, "Spinsters," 88; Nancy Cott, *The Bonds of Womanhood: 'Woman's Sphere' in New England, 1780–1835* (New Haven: Yale University Press, 1977), 57; Polly Welts Kaufman, *Women Teachers on the Frontier* (New Haven: Yale University Press, 1984), 33–34.

63. Classification of School Teachers of Over Twenty Years' Service, *SAPP* 1895, no. 105, 1; Leaves of Absence in the Education Department, *SAPP* 1900, no. 118, 1–8; see also Alison Mackinnon, "Awakening Women: Women, Higher Education and Family Formation in South Australia, 1880–1920" (PhD thesis, University of Adelaide, 1989), 78, 164–167; Alison Mackinnon, *One Foot on the Ladder: Origins and Outcomes of Girls' Secondary Education in South Australia* (St. Lucia: University of Queensland Press, 1984), 134–136.

64. Katie Holmes, "'Spinsters Indispensable': Feminists, Single Women and the Critique of Marriage, 1890–1920," *Australian Historical Studies,* no. 110 (1998): 74.

65. Catherine Helen Spence, "Some Social Aspects of South Australian Life," in *Catherine Helen Spence,* ed. Helen Thomson (St. Lucia: University of Queensland Press, 1987), 544.

66. Carol Bacchi, "The 'Woman Question,'" in *The Flinders History of South Australia: Social History,* ed. Eric Richards (Adelaide: Wakefield Press, 1986), 405–408; see also Haniford, "Crisis in Masculinity," 16.

67. Holmes, "Spinsters Indispensable," 76.

68. Megan McMurchy, Margot Oliver and Jeni Thornley, *For Love or Money: A Pictorial History of Women and Work in Australia* (Ringwood: Penguin, 1983), 42.

69. For further discussion, see Haniford, "Crisis in Masculinity," 13–23; Blount, "Spinsters," 88–92.

8
—

Transformations in Teaching: Toward a More Complex Model of Teacher Labor Markets in the United States, 1800–1850

Kim Tolley and Nancy Beadie

From the eighteenth to the late nineteenth century, school teaching in the United States transformed from a predominantly male to a predominantly female occupation. In an attempt to explain why education systems change over time, recent studies of education and state formation have focused on the interactions between the state and communities in the development of public schooling.[1] Unfortunately, this approach tends to overlook or ignore transformations in schooling that have occurred in the comparative absence of state intervention. In contrast, this chapter examines the shift from male to female teachers that occurred in two states from 1800 to 1850, a period that preceded the expansion of large public school systems in the United States.

This chapter investigates the role of the education marketplace in facilitating the entry of women into the occupation of teaching. In her comparative study of the social origins of educational systems in Russia, England, and other countries, Margaret S. Archer argued that "change occurs because new educational goals are pursued by those who have the power to modify previous practices."[2] Other scholars have positioned the locus of educational change in the interactions among subordinated and dominant social or political groups. What all of these theories share is a focus on social or political struggles rather than on supply and demand in the education market. While we do not seek to minimize the influence that that social and political struggles have had on the evolution of educational systems, in this chapter we argue that in the United States, supply and demand in the education market played a role in facilitating the access of women to teaching positions during the early national period.

* * *

This chapter brings together evidence from disparate local sources in North Carolina and New York to explore the structure of teacher labor markets from 1800 to 1850. Among the sources we examined are hundreds of newspaper advertisements for venture schools and academies in North Carolina, and detailed school and teacher employment records for one rural New York town. Considered in conjunction with scattered comparative material from other primary and secondary sources for those states, this evidence leads us to challenge a few key assumptions and suggest a few new lines of inquiry about regional variations in teacher employment, socioeconomic incentives for teachers, and the gender transformation in teaching in the early nineteenth century.

Specifically, this chapter presents evidence to suggest that during 1800–1850, the shift from male to female teachers was well under way in entrepreneurial schools in North Carolina. In fact, from 1820 to 1840, the rate of women entering academy and venture schoolrooms in North Carolina may have been higher than the rate of women taking up teaching in some areas of the Northeast. Similarly, it appears that in rural New York, the expansion of female teaching preceded, and occurred largely independent of, the tax-based system of support for common schools. In short, evidence from both places suggests that feminization occurred first through the market.

The Entry of Women into Teaching from 1800 to 1850

Over the past several decades, scholars have advanced a number of theories to explain the process that has been referred to as the "feminization of teaching." Factors believed to have motivated women to enter schoolrooms during these decades include a shortage of marriageable men, an increased demand for schooling as more towns established common schools, a low supply of men willing to teach, cultural assumptions about woman's sphere that provided ideological support for women teachers, and an evangelical commitment to missionary work.[3] Considering the question from the perspective of men's potential annual earnings, some scholars have claimed that men left teaching in the United States and Canada as the nature of schooling became more systematized. In particular, as the school year lengthened and as teaching evolved into "full-time" work, incompatible with other forms of employment, men accustomed to teaching the short "traditional" school term during the winter months and pursuing other lines of work during the summer left the profession.[4]

Although many researchers have focused on the post–Civil War era when discussing this shift in teaching, documentary sources indicate that in some areas of the Northeast, women began to teach in common schools and academies decades earlier. Massachusetts and New York school reports provide the earliest statewide data on the gender of teachers in common schools and academies. In both states, the proportion of women among teachers in common schools appears to have been expanding well before 1850. By 1829, women comprised 53 percent of teachers in Massachusetts's common schools when data for winter and summer sessions are combined, a figure that grew to 68 percent by 1847.[5] Similarly, in New York, women represented 62 percent of teachers in such schools by 1842 and 69 percent by mid-century.[6] Nor was women's entry into teaching restricted to the elementary levels. By 1857, women represented 50 percent of all the teachers in New York academies.[7]

To learn more about the gender of the teaching population in the early nineteenth century and in regions other than the Northeast, it is necessary to look at other kinds of sources. Common schools did not exist in North Carolina, until after 1840, when the legislature began to provide funding for such institutions.[8] Until then, formal schooling occurred through the education marketplace in entrepreneurial venture schools and academies. A venture school is defined here as an unincorporated institution, operated on an entrepreneurial basis, and supported entirely by tuition. An academy, by contrast, was a legally incorporated institution, governed by a board of trustees, and often partially subsidized by endowment income or other nontuition funding. Generally, both venture schools and academies in North Carolina served students between the ages of seven or eight and twenty-five.

By creating a database of teachers mentioned in North Carolina newspaper sources and other documents, it is possible to analyze the changing proportions of men and women by decade. North Carolina newspaper sources provide scattered information about the positions available in antebellum schools, the instructors who filled them, members of the boards of trustees, tuition rates, school sessions, commencement exercises, subjects offered for study, and other important details.

In the sample of 486 teachers analyzed for this study,[9] the proportion of women among teachers in North Carolina schools and academies appears to have increased steadily from 1800 to 1840, at a greater rate than that of men (see table 1). The largest increase occurred from 1810 to 1820, when the numbers of women mentioned in newspapers almost quadrupled while the numbers of men increased 67 percent. By 1830, the number of men had doubled while the number of women had grown 530 percent. The fourth decade witnessed a decided shift in the gender composition of this sample. From 1830 to 1840, the number of women continued to grow, whereas the number of men fell 27 percent, reaching a level lower than that of two decades earlier.

Forty-four of the teachers in this study are specifically identified as having come from the North, a figure that represents 9 percent of the total sample.[10] Just as in the larger sample, from 1810 to 1820, the number of females from the North increased dramatically, and from 1830 to 1840, the numbers of Northern men witnessed a decided drop, falling more than 90 percent. In contrast, the numbers of Northern women mentioned in North Carolina newspaper advertisements and other sources increased during the same period, as shown in table 1.

Table 1 Number and Percent of Teachers Appearing in North Carolina Newspapers and Other Documentary Sources, by Gender and Region of Origin, 1800–1840 (*n* = 486)

Decades	Total male	Northern male	Southern male	Total female	Northern female	Southern female
1800–1810	69	5	64	10	1	9
(n = 79)	(87%)	(7%)	(93%)	(13%)	(10%)	(90%)
1811–1820	115	9	106	38	6	32
(n = 153)	(75%)	(8%)	(92%)	(25%)	(16%)	(84%)
1821–1830	139	13	126	49	7	42
(n = 188)	(74%)	(9%)	(91%)	(26%)	(14%)	(86%)
1831–1840	102	1	101	63	12	51
(n = 165)	(62%)	(>1%)	(99%)	(38%)	(19%)	(81%)

Sources: Sample derived from all of the teachers mentioned in the following sources: Monthly interval sampling of the *Raleigh Register,* 1800–1840, North Carolina Collection, Wilson Library, University of North Carolina, Chapel Hill; Charles L. Coon, *North Carolina Schools and Academies 1790–1840: A Documentary History* (Raleigh: Edwards & Broughton, 1915); Mary Ellen Gadski, *The History of the New Bern Academy* (New Bern: Tryon Palace Commission, 1986), 166–168; Susan Nye Hutchison Diary, Southern Historical Collection (SHC), University of North Carolina, Chapel Hill; Ernest Haywood Papers, files 143–144, box 3, SHC; Mordecai Family Papers, files 1–11, box 1; file 113, box 8; file 15, box 2, SHC; John Steele Papers, files 67–69, box 4, SHC.
Note: Some teachers may be represented more than once if they taught across a span of two or more decades. Because they also taught at these institutions, school principals/heads are included in this sample.

Around 1830, both Northern and Southern men appear to have begun to leave academy and venture-school teaching in North Carolina. Not only did the migration of Northern men to the South taper off, but also the numbers of Southern men mentioned in documentary sources fell to levels below those of the period two decades earlier. From 1800 to 1840, the total number of male teachers rose only 48 percent, whereas the total number of female teachers increased more than fivefold. Some of this growth occurred because of a large increase in the numbers of Northern women coming South to teach, but even when the Northern women are removed from the sample, it is clear that the number of women teaching in North Carolina increased during the same period that the numbers of men fell, and that this trend began during the years from 1830 to 1840.

The same development can be seen in other kinds of documentary sources—the records of individual institutions. For a number of imaginable reasons, the proportions of men and women appearing in contemporary newspapers may not accurately represent the gender of the teachers in schools. It is possible that only the wealthier teachers paid for advertisements. It is also conceivable that women may have taken out more advertisements than men, particularly if they were trying to establish themselves in a male-dominated profession. Examining the extant records of individual institutions, which make it possible to identify all the teachers over an extended period of time, provides one means of checking the findings from newspapers against other sources. An analysis of the teachers in three schools in New Bern, Raleigh, and Warrenton reveals somewhat similar results.

Established as a coeducational academy, New Bern became the second school to receive a colonial charter following the incorporation of Philadelphia's Franklin Academy in 1753. Unlike the majority of antebellum schools that flourished for a few years and then disappeared, New Bern Academy remained in almost continuous operation from 1766 to 1882, in the sense that some form of schooling operated in the academy buildings throughout this period. Raleigh Academy, granted a charter in 1801 by the state legislature, endured for nearly three decades as a coeducational institution. Jacob Mordecai's entrepreneurial female school in Warrenton opened in 1809 and thrived for ten years, until the Mordecai family cashed out and left the state to take up farming in the West. All three schools enrolled over a 100 students each at some point in their history, and documentary sources provide information about the gender of the teachers working in these institutions at specific dates, as shown in table 2.

Reflecting the results obtained by analyzing newspaper advertisements and other documents, the records of these three schools indicate that the number and proportion of women among North Carolina teachers rose during the antebellum period. Given that some documentary sources may not have mentioned single women or women who worked as teaching assistants under more prominent men in coeducational institutions, it is possible that the percentage of female teachers during the antebellum period may have been actually somewhat higher than represented in table 2.

Taken as a whole, the documents analyzed in this study suggest that while the relative proportion of men and women in this sample of teachers was never static, it shifted more dramatically during specific decades. First, the greatest percentage increase among female teachers in North Carolina schools occurred

Table 2 Number and Percent of Male and Female Teachers in Three Antebellum North Carolina Schools

Years	New Bern Academy		Raleigh Academy		Mordecai's School	
	Males	*Females*	*Males*	*Females*	*Males*	*Females*
1776	1 (100%)	0	—	—	—	—
1804	1 (100%)	0	2 (67%)	1 (33%)	—	—
1809	1 (100%)	0	3 (60%)	2 (40%)	1 (100%)	0
1815	2 (67%)	1 (33%)	4 (67%)	2 (33%)	1 (33%)	2 (67%)
1827	*Insufficient Data*		2 (33%)	4 (67%)	—	—
1837	2 (67%)	1 (33%)	—	—	—	—
1844	2 (67%)	1 (33%)	—	—	—	—
1861	2 (40%)	3 (60%)	—	—	—	—
1871	1 (33%)	2 (67%)	—	—	—	—
1882	2 (25%)	6 (75%)	—	—	—	—

Sources: Data compiled from advertisements in the *Raleigh Register,* 1800–1830, North Carolina Collection, Wilson Library, University of North Carolina, Chapel Hill; Charles L. Coon, *North Carolina Schools and Academies 1790–1840: A Documentary History* (Raleigh: Edwards & Broughton, 1915); Mary Ellen Gadski, *The History of the New Bern Academy* (New Bern: Tryon Palace Commission, 1986), 166–168; Susan Nye Hutchison Diary, SHC, University of North Carolina, Chapel Hill; Mordecai Family Papers, SHC.

from 1810 to 1820. This decade also witnessed the first wave of Northern women migrating to the state to teach. Second, the numbers of Northern women mentioned in documentary sources rose again from 1831 to 1840, far outpacing the increase in Southern women and thus constituting a second wave of Northern migration. Third, during this decade, the migration of Northern men to the South ground nearly to a halt, and the overall number of men in schools and academies in the South fell for the first time.

To some extent, the increase in the proportion of female teachers can be explained as the result of an increase in the number of schools serving female students. The historian Charles L. Coon identified 121 institutions bearing the designation "academy" or "seminary" operating in North Carolina between 1800 in 1840. Of these, an increasing number served female students, either in single-sex or coeducational institutions. By the fourth decade, at least half of such institutions enrolled females.[11] At the highest levels of schooling available by mid-century, women's enrollment appears to have outstripped that of men. According to the first report of the state's superintendent for education, published in 1854, the enrollment of students at male colleges in that year was "perhaps between 500 and 600" whereas the "number at Female Colleges, (including Salem School and St. Mary's), [was] not less than 1,000."[12] However, the expansion of schooling for females does not entirely explain the increase in female teachers, because many of the earliest female and coeducational schools operated with predominantly male faculty, as illustrated by the examples in table 2.

Once we recognize that the entry of women into teaching began well before 1830 in both the North and the South, two long-standing theories about the feminization of teaching are called into question. First is the idea that traditional social norms proscribed women from becoming teachers in the early nineteenth century, and that these social norms were not overcome in the South until the post–Civil War period. Second is the theory that teaching became feminized with the bureaucratization of schooling and the extension of the school year. According to this view, teaching became less attractive to men who had taught only several months a year to supplement their income from other sources.

Social Norms

Scholars sometimes assume that in the early nineteenth century, traditional social norms proscribed women's participation in paid work, including teaching, and that these norms had to be overcome in order for feminization of teaching to occur. At the same time, a large body of scholarship demonstrates that women's participation in productive and paid labor extends well back into the colonial period, and that the ideological belief that daughters should be self-supporting can be traced at least back to the Revolutionary era.[13] In a recent survey of female advice literature published during the early national period, Margaret A. Nash found that "advocates of female education touted the practical benefits" of self-reliance and "self-sufficiency."[14]

The view that young women should contribute toward the support of their families was widespread in New England in the 1830s, as evidenced in the stories written by mill women in the *Lowell Offering*.[15] Nevertheless, Thomas Dublin's study of the letters written by female mill workers led him to conclude that daughters did not necessarily send their wages home, nor did all parents expect that they would. A young woman's economic independence was "useful" to the family, simply because her departure from the farm relieved the parents of the expense of supporting her. As Lucy Larcom wrote, when she left mill-work for teaching, "It had been impressed upon me that I must make myself useful in the world, and certainly one could be useful who could 'keep school' as Aunt Hannah did."[16]

Much of the scholarship on women's work in the late eighteenth and early nineteenth centuries focuses on New England, but evidence suggests that expectations regarding women's capacity for paid labor and self-support were more widespread than is often assumed. The presence of Southern women in North Carolina academies during the antebellum period suggests that in the South as well as the North, women violated no social norms when they established venture schools or began to teach in chartered academies. North Carolina's New Bern Academy educated girls in a female department when it opened in 1766, a development that caused no controversy in the community. When Raleigh Academy was established in the state capital in 1801, the school opened with a female department run by a woman.[17]

Positive social attitudes toward female teachers appear to have allowed the entry of women into schoolrooms in both regions of the country from the beginning of the nineteenth century. Although the overall proportion of women in teaching was smaller in the South than in the North, this study did not find any documentary evidence to support the argument that North Carolina residents opposed the efforts of females to teach in female schools or in the female departments of coeducational institutions.

The Bureaucratization of Schooling and Extension of the School Year

Several historians have identified the bureaucratization of schooling and the extension of the school year as a factor that induced men to leave teaching. According to this argument, men who traditionally had supplemented winter teaching with other occupations left teaching as the state instituted policies requiring longer school years, compliance with state certification procedures, and subordination to a hierarchy of school officials.[18] At first glance, this argument is compelling when considered from the perspective of common school teachers in the Northeast. In Massachusetts and New York, the timing of increased female participation in teaching roughly corresponded to the timing of state intervention in common schooling and the lengthening of the school year, suggesting that the very fact of state involvement may have created the conditions that fostered feminization. A close look at employment practices at the local level, however, challenges this idea.

In the town of Lima, New York, the lengthening of the school year and the expansion of female teaching developed together, but largely outside the structure of tax-supported schooling. Located in the agriculturally rich region of western New York known as the Genesee Valley, Lima was settled by New Englanders in the post-Revolutionary era and remained rural throughout the nineteenth century.[19] As in Massachusetts, town and district schools in rural New York typically hired male teachers for winter schools and female teachers for summer schools. In Lima, this practice was in place from the very beginning of the state-administered system of tax support for common schooling in 1815. It was not until 1830, however, that school district leaders in Lima allotted any portion of its tax funds to support summer schools. Even after the introduction of some tax-based subsidies for summer school in 1830, more than three quarters of summer school costs continued to be financed by tuition.[20]

Despite, or perhaps because of, this lack of tax support for summer schooling, the length of the summer term expanded over the course of the 1820s, from three to four to five to six months a year. Meanwhile, the length of the winter term stayed the same, stabilizing at four months. The extent of female teaching and the total length of the school year expanded, in other words, but without any direct influence from the state. Rather, the feminization of teaching and the expansion of the school year in Lima developed first as part of a market- or tuition-based approach to schooling, which a tax-based state system gradually absorbed.[21]

To some extent, this interpretation complements the conclusions of Joel Perlmann and Robert Margo. They also reject the theory that feminization of teaching occurred in response to the bureaucratization of schooling and lengthening of the school year. Instead, they adopt the "simpler" but related theory that feminization developed in response to school boards' desire to hire cheaper labor.[22] In their analysis, feminization developed first in New England because a distinctive two-tier system of female summer schools and male winter schools had existed there since the colonial era, making it easier for women to make the transition to year-round teaching, and because the gap between male and female wages was much greater there than elsewhere. To the extent that the data from Lima suggest that market forces drove feminization, they are consistent with certain aspects of Perlmann and Margo's conclusions.[23]

In constructing their account, however, Perlmann and Margo focus almost exclusively on common school teaching, thereby missing evidence of feminization in the South, such as that presented here for North Carolina. North Carolina schools, like New York academies, typically hired both male and female teachers for annual terms. As shown in table 1, the numbers of female teachers increased and the proportion of male teachers declined in an educational environment characterized by a lack of state-supported common schools and a norm of year-round teaching in academies and venture schools. Contrary to Perlmann and Margo's assumptions, then, our evidence suggests that feminization also occurred in the absence of a two-tier system of female summer schools and male winter schools, and in the absence of a dramatic gap between male and female wages.

Moreover, Perlmann and Margo analyze the influence of labor markets and the issue of feminization almost entirely from the district viewpoint. In their analysis, the question is when the male-female wage differential became great enough to make it financially worthwhile for districts to overcome the sanctions against hiring female teachers. This focus ignores those sectors of schooling in which the initiative for organizing schools and establishing rates lay largely with teachers themselves, that is, venture or market-based schooling. When viewed from the perspective of teachers, rather than districts, feminization may have been driven as much by incentives for women to enter teaching as by incentives for districts to hire them. What is more, this perspective suggests the possibility that women effectively created a market for their services that later gave them the leverage to negotiate a place in state-based systems. This possibility leads us to look more closely at the structures of opportunity for female teachers from the perspective of women themselves, and to consider opportunities outside as well as inside tax-based systems of common schooling.

Economic Incentives for Women to Enter Teaching, 1800–1850

In recent decades, a significant body of scholarship by labor historians and scholars interested in women's history has revealed much about the transition of women workers from the home and local community to small manufacturing centers and large factories in the early nineteenth century.[24] However, what is still largely missing from the secondary literature is a serious consideration of women's work as teachers during the early antebellum period. Most published discussions of antebellum teaching salaries have characterized teachers' wages as comparable with those paid to domestic servants.[25]

How did the wages paid to New England common school teachers compare with the wages a woman could earn by doing other kinds of work? According to Alice Kessler Harris, Philadelphia seamstresses in 1821 made about $1.05 per week, out of which they paid for thread, heat, light, and room and board. A shoe binder in the 1820s could earn from 72 cents to $2 per week and a hatmaker might earn from $1.50 to $1.75 per week. Like seamstresses, both shoe binders and hatmakers had to pay board out of their earnings, unless they worked at home.[26] In Lima, female common school teachers earned wages that fluctuated between $4 and $6 per month in the 1820s and early 1830s, with board as an additional benefit (see table 3).

Adding the value of board to the wages of these female teachers brings the value of their earnings to $7 to $9 per month, or $1.69 to $1.93 per week, earnings that are 10 percent to 13 percent higher than the earnings of hatmakers, 60–85 percent higher than that of seamstresses, and more than double that of some shoe binders. This suggests that for women, common school teaching could be attractive as compared with other forms of wage work. Female common school teachers in rural towns such as Lima could earn such wages for a maximum of five to six months a year, however, while shoe binders and seamstresses (though perhaps not hatmakers) presumably could work longer. Whether such alternative employment existed for any particular female, of course,

Table 3 Adjusted Wage Rates of Female Summer School Teachers School District #4, Lima, New York, 1820–1833

Year	Index	Monthly wage	Adjusted wage	Board	Wage plus board	Adjusted total
1820	141	$5.95	$4.22	$3.19	$9.13	$6.47
1825	119	4.00	3.36	3.19	7.19	6.04
1830	111	5.00	4.50	3.19	8.19	7.37
1833	101	4.00	3.96	4.25	8.25	8.16

Sources: Data compiled from *Record Book, Lima School District #4, 1814–1854,* Lima Historical Society, Tenny Burton Museum, Lima, New York. The index used to convert wages to constant value terms is the Composite Consumer Price Index from John J. McCusker, *How Much is That in Real Money? A Historical Price Index for Use as a Deflator of Money Values in the Economy of the United States* (Worcester: American Antiquarian Society, 1992).

depended on the location and her geographic mobility. For women in rural areas outside the major textile manufacturing regions of New England and the Upper Hudson Valley, school teaching may well have been one of the few available options for wage work.

Although there existed nothing really comparable to the New England common school in early national North Carolina, documentary evidence indicates that in Raleigh, at least, a teacher could earn a comparatively high salary teaching the rudiments to poor children. Although schooling in North Carolina was generally entrepreneurial and market-driven, many communities and religious groups organized and subsidized various forms of charitable schooling. In 1822, the Raleigh Female Benevolent Society paid a teacher to teach reading, writing, arithmetic, and "all kinds of plain work" to roughly 26 poor female students during the week and to provide instruction to a larger group of students on Sunday. It is possible that the teacher also received free room and board, because the society offered board when it advertised to fill an opening in the position five years later.[27] The society paid its teacher an annual salary of $200 in 1822. In monthly terms, this represents a wage of $16.66 per month, more than double the monthly wage of common school teachers in rural western New York. To understand how this might compare to other kinds of wage work, it is instructive to examine Thomas Dublin's data of the overall earnings of palm-leaf hatmakers in Fitzwilliam, New Hampshire (table 4). According to Dublin, young women making hats in Fitzwilliam "rarely worked at it full time." In analyzing the overall wages from hat making, he found that eight Fitzwilliam families earned an average total amount of $220 each over an 18-year period. Comparing the hat-making earnings to the wages of mill workers, Dublin concluded, "It probably would have taken a teenage daughter four or five years of mill employment to have saved this much money to contribute toward her family's expenses."[28] From this perspective, the $200 annual salary offered in Raleigh appears very attractive. Moreover, the position in Raleigh offered annual, rather than seasonal, employment.

During the 1830s, Massachusetts began to collect systematic data on its common school teachers, allowing some comparison of teachers' weekly earnings with that of other workers. For the sake of comparison, table 4 incorporates wage data from both primary and secondary sources. This data shows that female common school teachers earned substantially more per month than weavers, hatmakers, paper mill workers, and some textile workers, and slightly more than Lowell textile operatives, though it is not entirely clear how comparable the terms and length of employment were.

Because economic historians have studied labor income by industrial sectors in the 1840s, we can use previously published data to compare the income of common school teachers in New York, Massachusetts, and Connecticut with the income of workers in agriculture, manufacturing, and all other industrial sectors. Making such a comparison involves converting all wages to constant value terms, as shown in table 5.

Together, these various wage comparisons for the period of the 1820s through the 1840s suggest that common school teaching offered wages exceeding those offered for other forms of work available to women, with the possible exception of manufacturing in the mid-1840s.[29] Variations in the terms of employment (whether board is included), across rural and urban locations, and in the number of months of employment available for different kinds of work make comparisons difficult. Nonetheless, the economic incentives for women to enter these forms of teaching appear to have been real.

However, common school teaching was not the only form of teaching available to women in the early antebellum era. In both the North and the South, academies and venture schools offered teaching opportunities beyond the common school. Academy teaching presented several advantages over common school teaching, and at least one advantage over operating a venture school. First, academies offered annual salaries and employment. In both New York and North Carolina, a typical academy schedule consisted of two five-month sessions

Table 4 The Weekly Earnings of Selected Women's Occupations in Massachusetts and New Hampshire, 1822–1837

Sector	Weekly wages
Weavers, Richmond, NH (1822–1829)	$0.42
Palm-leaf hatmakers, Fitzwilliam, NH (1830)	$0.34
Paper mill workers, South Hadley, MA (1832)	$2.65
Textile workers, Chicopee, MA (1832)	$2.75
Lowell female textile operatives (1836)	$3.25
Lowell female common school teachers (1837)	$3.50

Sources: Data compiled from Thomas Dublin, *Farm to Factory: Women's Letters, 1830–1860* (New York: Columbia University Press, 1993), 11–12; Dublin, *Women at Work* (New York: Columbia University Press, 1979), 161; Dublin, *Transforming Women's Work: New England Lives in the Industrial Revolution* (Ithaca: Cornell University Press, 1994), 41, 59; *Abstract of the Massachusetts School Returns for 1837* (Boston: Dutton & Wentworth, 1838), 50.

Note: The value of the board provided to paper mill workers has been calculated and added to their reported weekly wages for the sake of comparison to shoe binders and hatmakers, who had to pay room and board out of their earnings. Board at Chicopee was valued at about one-third of women's earnings; this rate was applied to the wages of paper mill workers.

Table 5 Average Monthly Wages of Female Common School Teachers in Three Northeastern States, and Labor Income of Female Workers in Agriculture, Manufacturing, and All Other, Reduced to Constant Value Terms and Expressed in 1845 Dollars

Sector	Monthly income per worker, in 1845 terms	Value of wages with board factored in
Agriculture	$ 5.25	—
Manufacturing	$10.00	—
All Other	$ 7.95	—
Connecticut common school teachers, 1846	$ 6.83	$ 9.67
New York common school teachers, 1845	$ 7.00	$ 9.92
Massachusetts common school teachers, 1847	$12.50	$ 12.50

Sources: Eleventh Annual Report of the Board of Education (Boston, Massachusetts, 1848), 26; *Annual Report of the Superintendent of Common Schools of Connecticut* (Hartford: Case, Tiffany, and Burnham, 1846), 8; *Annual Report of the Superintendent of Common Schools of the State of New York* (Albany: Carroll and Cook, 1845), 13; "Labor Income Per Worker, by Industrial Sectors, 1840," in Lance E. Davis, Richard A. Easterlin, William N. Parker et al., *American Economic Growth: An Economist's History of the United States* (New York: Harper & Row, 1972), 26.
Note: The labor income for agriculture, manufacturing, and "all other" is defined as gross income less gross property income. This figure has not been adjusted to allow for the costs or value of board. The table uses a female to male wage ratio of 0.45 to approximate female wages for agriculture, manufacturing, and "all other."

or four quarters. For adult women who could no longer rely on parental support in the off-season, an academy position provided a more viable means of support for self and family support than teaching common school, which did not offer annual employment.

Second, chartered academies generally offered more security than venture schools. An academy was a corporate institution that held title to some property and operated under the authority of a board of trustees. This property and corporate governance provided some cushion against the vagaries of the market. Although academies, like venture schools, depended heavily on tuition for income and salaries, their corporate property and status gave them a degree of financial and legal security that an individual venture schoolteacher could not enjoy. Moreover, the social prominence and networks of academy trustees gave them greater leverage in recruiting students, collecting student fees, and raising capital. In New York after 1816 and in North Carolina after 1840, chartered academies also received some subsidies from state-endowed funds, providing them further financial security. Trustee governance did entail some loss of autonomy on the part of the teacher, and may also have involved some loss of earning potential when the market was good. Nonetheless, academies offered what may have been considered a reasonable trade-off for adult teachers seeking a stable income.[30]

Third, academies paid more than common schools. The opening of Genesee Wesleyan Seminary, a coeducational academy in the town of Lima, in 1832, provides a rare opportunity to compare directly the opportunities and rewards for common school and academy teachers in the same place and time, for both males and females.[31]

As shown in table 6, Genesee Wesleyan Seminary paid higher salaries to its three female teachers than women earned in Lima's district schools.

These differences between the salaries of common school teachers and academy assistants, and between assistants and female heads of academy departments, suggest a fourth advantage of academy teaching. It offered opportunities for career and social advancement. For New York men and women both, common school teaching in an ordinary rural school was the bottom of the ladder with regard to salary and terms of employment. It was, nonetheless, a place where they could begin acquiring the experience, and perhaps the money, that could enable them to move up the ladder.

For women in particular, the opportunity to become an academy teacher, the head of a female academy department, or the principal of an all-female school may well have provided an incentive for entering teaching in the first place, even at the relatively low wages paid by rural common schools. Few, if any, other lines of work offered women the possibility of higher positions and salary improvement over time.

The gender gap in wages narrowed as female teachers climbed the career ladder, though within limits defined by their sex. Among common school teachers, the female/male wage ratio consistently hovered around 0.43 through the 1820s and 1830s, virtually the same ratio that Perlmann and Margo found for male and female laborers in antebellum New York.[32] In contrast, among academy teachers of equivalent status, the female/male wage ratio was much higher,

Table 6 Local Structures of Opportunity: Common School and Academy Teachers Lima, New York, 1833

Sex	Common School		Academy			
	Monthly wage	Monthly wage plus board	Annual salary without board			
	Summer	Total per month/board	Assist. primary dept.	Music	Female dept. head	School princ./ head
Female	$ 4	$ 8.25	$100–$150	$200	$200	
	Winter	Total per month/board			English dept. head	Lang. teacher
Male	$15	$19.25			$300	$375	$500
Female/ Male Ratio	.27	.43			.67		

Sources: Data compiled from *Record Book, Lima School District # 4, 1814–1854*, Lima Historical Society, Tenny Burton Museum, Lima, New York, and from *Account Book #178, Journal of the Doings of the Legal Board of Trustees of the Genesee Wesleyan Seminary, 1830–1854*, Genesee Wesleyan Seminary Collection, Archives and Special Collections, Byrd Library, Syracuse University, Syracuse, New York.

standing at 0.67 in 1833. This figure narrowed over time and according to position, ranging between 0.71 (for assistants) and 0.84 (for heads of departments) in the late 1840s.[33] These relatively high female-to-male wage ratios in turn raise some interesting questions about teacher labor markets in the antebellum era. They suggest that at certain levels at least, women exercised market leverage. In some sectors and places, in other words, the demand for female teachers approached or exceeded supply. What factors shaped this supply and demand?

The first thing to understand about the position of female teachers in academies is that they were regarded as necessary for most coed schools. Filling the position of "preceptress," or head of the female department, with a respectable and effective female teacher made a significant difference to the financial health and viability of an institution. Without such a person, an academy could not effectively attract female students, especially older female students, and without such students, academies lost an important source of income. As Beadie has established elsewhere, the capacity to attract female students was an important source of financial success among academies in New York.[34] This importance increased during the antebellum era, as the number of female academy students in upper-level subjects statewide came to equal and then exceed that of males in the late 1840s. The enrollment of females could not only increase the size of an institution's pool of potential students, but it could also have a financial impact out of proportion to simple numbers. Girls and young women were by far the largest clientele for subjects such as music and drawing, which academies offered for extra fees. These fees could be substantial, as much as two to four times those charged for regular academic subjects. In addition, there is some evidence that women were more likely to persist through a full academy course. But the importance of female teachers was not limited to their influence on female students. In addition, female teachers superintended the primary departments of many coeducational schools that offered basic English instruction to younger children of both sexes. Throughout the antebellum era, one-third to one-half or more of all students attending any particular academy enrolled in these common school subjects.[35] In all these ways, then, female teachers were essential to the operation and finances of academies like Genesee Wesleyan.

Toward a More Complex Model of Antebellum Teacher Labor Markets: Sources of Variation

The case of Lima, New York, and Genesee Wesleyan Seminary provides a starting point for hypothesizing a model of teacher labor markets and opportunity structures in the antebellum era. To some extent, the salary pattern at Genesee Wesleyan reflected the broader trends among state-subsidized academies in New York. In a separate study of state-level academy data, Beadie found that the salary range for academy teachers narrowed over time across the state. Between 1840 and 1850, for example, the lowest average teacher salary paid by any institution rose, while the highest average salary paid by any institution declined.

During the same time period, similar trends occurred in tuition pricing, a convergence that suggests a competitive market had emerged among academies.[36]

Overall, this convergence had the effect of lowering average teacher salaries for the system. When viewed in relation to the particular example of Genesee Wesleyan, however, a gender interpretation of this decline in salaries can be made. As teacher labor markets developed, the salaries of male teachers at the top of the hierarchy may have declined at the same time as salaries of female teachers at the bottom of the hierarchy rose. If so, such a trend could contribute substantially to understanding the feminization of teaching. It could help explain why men, who experienced a decline in opportunity, left the profession, even as women, who experienced an improvement in opportunity, entered teaching in greater numbers.

Sources for North Carolina illustrate how the pressures of economic depression and competition among schools could open up opportunities for increased female participation in teaching. During times of recession, schools lowered the price of tuition in order to stay in business. The years from 1820 to 1821 were particularly difficult for educators in North Carolina. The financial Panic of 1819 delivered a harsh blow to the Southern economy. In 1820, the Raleigh Private Academy slashed its tuition to match that of a competitor in town. In 1821, Warrenton Female Academy reduced its terms by 20 percent. In the same year, Shocco Female Academy announced that it would maintain its relatively low rate of board and tuition, stating, "While times continue as they are, the price of Board and Tuition will be one hundred dollars per annum . . . payable in advance."[37] During the 1830s, when the number and proportion of male teachers fell in North Carolina, newspaper articles and advertisements indicate that some schools took steps to reduce their costs.[38] The Panic of 1837 ushered in the deepest and most prolonged depression in the antebellum period. In that year, the Episcopal School in Raleigh slashed its tuition "to meet the changes of times and the expectations of the Public."[39] In an environment of periodic economic recession and increased competition among venture schools, the opportunities for women to teach increased. A woman entrepreneur running her own school could undercut the terms charged by her male competitors and gain market share by attracting more students. Similarly, a board of trustees in a chartered coeducational school could lower costs by replacing a male assistant with a woman. Such replacements occurred first in subjects that had long been associated with female education, such as music, French, geography, and English grammar. By 1840, increasing numbers of women also taught the sciences and higher mathematics, and eventually Latin.[40]

Schools in the North similarly experienced the effects of economic depression. These effects are apparent in the simultaneous convergence in tuition and salary prices in the systemwide data for New York State.[41] In the decade after the Panic of 1837, New York Regents academies lowered their tuition in order to attract students, and they reduced teacher salaries in order cut costs. Added to the pressures of economic recession were those of increased competition among schools. Between 1838 and 1848, the number of Regents academies more than doubled statewide. Unlike entrepreneurial schools, Regents academies received some annual operating funds from the state, of which the total amount increased in 1838. However, the increased number of Regents institutions and of students attending them meant that the share of per-pupil state funding earned by most individual institutions declined in the 1840s.[42] Of course, for schools operating in North Carolina and elsewhere without any form of state support, the percentage of revenue derived from tuition would have been higher. During periods of economic recession, the easiest way for a school to cut costs involved either freezing or cutting salaries.

Incidental salary information from other institutions in New York State illustrates the differential impact such trends may have had on male and female salaries. This information shows how an overall decline in teacher salaries and a convergence in wages and prices could obscure an improving, or at least a stable, salary market for women. At Sherburne Academy in 1842, for example, trustees guaranteed their male principal a salary of $500 a year. For their female teacher, they specified a minimum salary of $250.[43] Teacher salaries thus averaged $375 per year. Several years later, at an institution known as Falley Seminary, in Fulton, New York, trustees also decided to pay the female "preceptress" of their "female department" $250 per year, but to pay their male teacher just $400 a year. The average teacher salary at Falley, then, was $325, or $50 less than that at Sherburne, though the salary of the female teacher was the same.[44]

Further comparative research into female teachers' salaries in both urban and rural locations, and in both northern and southern regions, is clearly needed before the full range and variation of opportunities for teachers can be understood. Data from North Carolina are suggestive, however, of the possibilities for female teachers with the education, experience, executive ability, social background, and mobility necessary to search out the most lucrative positions available in academies and venture schools (table 7). As early as the 1820s and 1830s, some North Carolina institutions advertised salaries as high as $500 for a female academy teacher or principal, figures that match those paid to male principals of coed academies in New York around the same time.

Whether the high salaries advertised for some positions in North Carolina reflect regional differences in wage markets for the South or the status of positions at the very top of the career ladder cannot be concluded from the data we have. Documentary evidence suggests, however, that Southern schools offered higher wages to female teachers than they could expect to receive in the North. In his 1847 report as superintendent of the Massachusetts Board of Education, Horace Mann indicated that Southern schools were advertising high salaries in order to attract Northern teachers.

"I regret exceedingly that I have not kept an account of the number of applications which I have received for the last ten years from the Southern and Southwestern states, for talented and highly qualified females, to take charge of select schools . . . at times, certainly, they have been as frequent as once a week, for a considerable period . . . The compensation offered varies from $400 to $600 a year—sometimes, also, including the journey to the place of employment. The average may be set down at $500. Many of the most highly educated young women of New England yield to these inducements."[45]

Table 7 Wages for Academy Teachers as Advertised by Eleven Higher Schools in North Carolina Newspapers, 1808–1841

Year	Male wage	Female wage
1808	$220.00	$125.00
1819	$500.00	$285.00
1822*	$1,361.00	$776.00
1824	$500.00–$600.00	$285.00–$342.00
1826*	—	$500
1829	$300.00–$400.00	$171.00–$228.00
1830*	$800.00	$456.00
1830	$300.00–$500.00	$171.00–$285.00
1837*	$900.00	$513.00
1839	—	$500.00
1841	—	$500.00

Source: The data is derived from newspaper advertisements included in Charles L. Coon, ed., *North Carolina Schools and Academies 1790–1840* (Raleigh: Edwards & Broughton, 1915), 206, 222, 803, 807, 811–815, 818, 820. *These schools wanted a teacher who would also serve as the principal/head of the school.

Similarly, Perlmann and Margo concluded on the basis of scattered sources from 1860 that female teachers in the South typically earned as much as 85–100 percent of the wages paid to male teachers.[46] Despite such economic incentives, a larger proportion of teachers were female in the North than in the South. In 1829, women represented slightly more than half of the common school teachers in Massachusetts, whereas women comprised only 26 percent of schoolteachers in North Carolina the following year. Although the rate of feminization in North Carolina schools may have been higher than the rate in Massachusetts from 1821 to 1840, a period that allowed North Carolina women to catch up, overall differences in the proportion of women in the teaching population persisted owing to developments prior to 1829. Conclusions about why this was the case must be tentative. Nonetheless, an existing body of secondary literature suggests a number of economic phenomena that may have contributed to regional differences in the supply and demand of female teachers during the early national period.

Differences in Regional Economies

Although the economic and technological developments of the early nineteenth century bore directly on the profitability of women's work in the home, these developments affected the northern and southern regions of the United States in different ways. A number of scholars have noted that the women who entered New England schoolrooms during the early nineteenth century hailed largely from farm families. Three developments occurred during the decade from 1810 to 1820 had a decided impact on the earning ability of women living on New England farms. First, the fledgling New England economy went from boom to bust during this decade, ushering in a period of economic depression in the region, running roughly from 1815 to 1822.[47] Second, technological improvements in textile manufacturing, coupled with intense competition from foreign markets, decimated the home-weaving industry, a traditional source of earnings for women. Third, a 15-year period of deflation in commodity prices originated in this decade, a phenomenon that put pressure on New England farm families to increase their earning power.[48]

The early antebellum depression and deflationary period did not affect all regions of the United States equally. While the Northeast experienced an economic depression and years of prolonged deflation, the South witnessed a brief cotton boom from 1815 to 1819. Although the Panic of 1819 dealt a blow to the southern economy, the region experienced a slow recovery because the prices of rice and sugarcane remained relatively stable and cotton production expanded enormously in response to increasing demand.[49] This is not to say that the crisis in textile manufacturing did not affect women in the South. In fact, the records of the Raleigh Female Benevolent Society indicate that indigent Southern women who relied on the loom and spinning wheel to make ends meet faced the same falling demand and lower prices as did their Northern sisters.[50]

Despite evidence that poor Southern women could no longer count on earning such a living from home textile work because the South had never heavily invested in textile manufacturing, the region largely avoided the broad social consequences of the late eighteenth-century boom and its consequent bust. As a result, Southern women living on farms or plantations may not have felt the same pressure as their counterparts in the Northeast to provide supplementary income to their families or support themselves through teaching.

Regional Differences in Access and Entry to Teaching

In addition to regional economic trends, differences in levels of access to entry-level positions in the occupation may have contributed to variations in the supply of female teachers in the North and South. In contrast to an occupation such as weaving or hatmaking, teaching required a long-term investment in education in order to gain the knowledge and skills necessary to teach, especially at the most lucrative levels. This requirement served as a powerful entry barrier to young men and women whose families could not afford to forgo their labor and send them to school. The widespread existence of district- and town-based common schools in the North not only facilitated access to the first rungs of this career ladder, but also fueled demand for teachers to fill common school positions as the population expanded. In contrast, teaching was far less accessible as a form of work for women from poor to middling families in the South. As we noted earlier, unlike New York, where some form of common schooling existed by 1820, North Carolina did not establish legislation in support of common schools until around 1840. Moreover, North Carolina was an exception among Southern states in making such provisions in the antebellum era.

The absence of a town- or district-based tradition of common school organization may have affected the supply of female teachers in the South in another way as well. In Perlmann and Margo's analysis, the roots of feminization of common school teaching in New England lay in the tradition of dame schools dating from the colonial era. A dame school offered rudimentary instruction in reading, writing, and arithmetic to children from two and a half to seven years of age. Although many women conducted entrepreneurial dame schools, some towns in New England states provided public funds to support these institutions.

Scholars have noted the absence of so-called dame schools in the South. Perlmann and Margo speculate that the absence of dame schools in the South may have been due to the fact that Southern women were more in demand in agriculture in the South than in the North, and that Southern women were less likely to have received the necessary schooling to teach in such schools.[51] The documentary sources examined for this study suggest another possible factor as well. It was very difficult to make a living teaching the so-called rudiments to very young children in a free-market context. In the absence of a tradition of town- or district-based common schools, North Carolina women with the background and interest in teaching had little economic incentive to organize schools for young children. North Carolina entrepreneurial schools competed for students able to pay high rates of tuition. Schools charged the highest rates for the most specialized branches of instruction: music, the classics, the higher mathematics, and other subjects that comprised the "higher schooling" expected of an academy or seminary. In contrast, the tuition that institutions could charge for teaching the alphabet and simple arithmetic to children younger than eight was very low.[52] In fact, it was so low that the only teachers to offer such instruction usually did so in preparatory classes of young children that studied alongside their more advanced peers. In this sort of arrangement, the provision of such preparatory classes ensured a pipeline of tuition-paying students capable of eventually pursuing higher, more profitable levels of study.[53]

Conclusion

In venture schools and academies, in the South as well as the North, women entered teaching in significant number as early as the 1810s, with increasing frequency in the 1820s and 1830s, and with apparent approval from leading citizens. In fact, the documentary evidence presented here suggests that from 1800 to 1840, the rate of women entering entrepreneurial schoolrooms in North Carolina may have been as high as the rate of women taking up teaching in some northern areas. For example, contemporary board of education reports for Massachusetts indicate that the proportion of female teachers in common schools increased 28 percent from 1829 to 1847. In North Carolina, the proportion of female teachers in venture schools and academies increased 46 percent during a comparable period, from 1821 to 1840. These figures indicate that women in both regions were entering schoolrooms during the same period that men were leaving the occupation.

This study contributes to our understanding of the transformation in school teaching by documenting the socioeconomic incentives for women to teach. In this study we establish the existence of informal career ladders for antebellum female teachers and begin to outline the structure of teacher labor markets across different levels and types of schools, including venture schools, common schools, and academies. In addition, we identify possible sources of variation in the salaries and opportunities available to women at different institutions and in different regions.

Our findings suggest that the feminization of teaching looks quite different when viewed from the teacher's perspective than it does when viewed from that of male school trustees. Existing studies of nineteenth-century teacher labor markets treat schools and school districts as the locus of decision making. From that viewpoint, the process of feminization turns on the question of when school trustees decided that the financial advantage of lower female wages made it worthwhile to hire female teachers. In this article, we examine the socioeconomic incentives for women to enter teaching from the perspective of teachers themselves, taking into consideration female teacher wages and opportunities in relation to other wages and forms of work available to women at the time. From this angle we discover not only that teacher wages compared favorably with other forms of paid work available to women, but also that women may have initiated their entry into teaching by organizing their own market-based venture schools, undercutting the schools of male competitors with lower tuition prices, and effectively creating their own teacher labor market. We also discover that opportunities for female career advancement and salary improvement existed, particularly in well-established academies in both the North and the South.

These findings contribute to the possibility of constructing a more complex model of teacher labor markets in the antebellum era. Most studies of teacher wages and feminization of teaching in the nineteenth century focus exclusively on common school teaching and analyze the entire issue of market and opportunity for female teachers as though it were limited to the wages offered at the bottom of the scale. It is no doubt true that well-paid female academy teachers represented a small portion of the female teaching force and, therefore, would not significantly change average wages and salary ratios for all female teachers as a group. Nevertheless, we would argue that academy teaching was significant for structuring female participation in the field. The *prospect* of relative financial independence and comparatively good salaries offered academy teachers might still have motivated women to prepare for and enter teaching, even if only a small proportion of such women actually held higher positions at any one time. Becoming the principal of a female academy was the apex of the career ladder for a female teacher, and may have been as close to equity a woman could get in any field or occupation in antebellum society.

This study also complicates our understanding of the interaction of the state, markets, and access to education during the antebellum period. In the case of Lima, New York, women's employment as summer school teachers occurred from the very first year of implementation of the state's common school laws and may well have predated the first year of their implementation in 1815. This initial expansion was driven more directly by market forces than by the state, as summer schools and female schoolteachers continued to be supported entirely through tuition rather than tax funds until 1830, and continued to be supported mostly by tuition for some period thereafter. The historian Kathryn Kish Sklar has found similar patterns of tuition-based support for summer schools and female teachers in local school records for towns in Massachusetts.[54]

At the same time, the tradition of town- or district-based schools in the North, however they were funded, may have promoted women's entry into teaching by providing access to the first rungs of the career ladder. One way in which Northern common school systems may have facilitated this access was by making it cheaper for women to receive a basic education than it was under the more fully entrepreneurial system prevalent in the South. As sources for North Carolina reveal, in order for an adult female teacher to be fully self-supporting, she needed to enroll older as well as younger students, teach higher as well as lower branches, charge more advanced students higher rates of tuition, and remain employed for most of the year. Northern common schools, by contrast, hired young women to teach young children basic subjects for four to six months a year, at relatively low wages. This arrangement may have made it less expensive and thus more common for women to receive an education and gain experience as teachers in the North. By the same token, however, it made it more difficult for women to earn a living as a teacher as an adult. To continue teaching over the long term, a woman had to leave common schools and find a position in an academy or a market able to sustain a successful venture school.

Much more research remains to be done. We need a clearer picture of the teachers in dame schools, venture schools, charity schools, and academies, in every geographic region of the country. More information about the gender shift in teaching during the antebellum period would give us a greater sense of the continuity between this earlier period and the late nineteenth century, an era where most scholars of feminization have focused their efforts. An investigation of the salaries paid teachers at institutions serving African American and Native American populations is also essential for establishing the range of opportunities available for female teachers, including both white women and women of color.

All these areas of research require that scholars make use of new sources of evidence. By focusing on systematic state and federal level data, historians have provided important information about gender and salary norms in certain sectors of education, periods of history, and regions of the country. Exclusive reliance on such data, however, overstates the significance of tax-subsidized common schools in the range of education and teaching opportunities in the country as a whole. It also obscures the interaction between different sectors of education and different regions of the country that shaped the structures of opportunity and career decisions of individual female teachers.

In conclusion, this study demonstrates that a transformation in the teaching populations of New York and North Carolina developed, to varying degrees, in response to supply and demand in the education market rather than as a result of direct state intervention. Socioeconomic incentives, coupled with a rising demand for female education, increased the proportion of women among teachers. It is not our intention to suggest that the influence of the market is more important than the influence of political initiatives, policy struggles, or what Ting-Hong Wong calls "the dialectical connection between education and state formation."[55] Nevertheless, as this case illustrates, the market can stimulate change in educational practice. By the time that large state-funded education systems developed in the later nineteenth century, the question of whether women would teach in American public schoolrooms had already been resolved. In many cases, the state simply appropriated the structures and processes developed in community-based, private, and voluntary schools.

Notes

1. To date, the most comprehensive treatments remain those of Margaret S. Archer and Andy Green. See Archer, *Social Origins of Educational Systems* (London: Sage Publications, 1979); Green, *Education and State Formation: The Rise of Education Systems in England, France and the USA* (New York: St. Martin's Press, 1990). For additional studies that examine the relation between education and state formation, see Bruce Curtis, *True Government by Choice Men? Inspection, Education, and State Formation in Canada West* (Toronto: University of Toronto Press, 1992); Stephen L. Harp, *Learning to be Loyal: Primary Schooling in Nation Building in Alsace and Lorraine, 1850–1940* (Dekalb: Northern Illinois University Press, 1998); Ting-Hong Wong, *Hegemonies Compared: State Formation and Chinese School Politics in Postwar Singapore and Hong Kong* (New York: Routledge, 2002).

2. Archer, Social Origins of Educational Systems, 2.

3. See John L. Rury, "Who Became Teachers and Why: The Social Characteristics of Teachers in American History," in *American Teachers: Histories of a Profession at Work,* ed. Donald Warren (New York: Macmillan, 1989), 9–48; Geraldine Jonçich Clifford, "Man/Woman/Teacher: Gender, Family and Career in American Educational History," in Warren, *American Teachers,* 293–343; Michael W. Apple, "Teaching and 'Women's Work'": A Comparative Historical and Ideological Analysis," *Teachers College Record* 86 (Spring 1985): 457–473; Kathryn Kish Sklar, *Catharine Beecher: A Study in American Domesticity* (New York and London: W. W. Norton, 1976), 97–98; Barbara Miller Solomon, *In the Company o f Educated Women* (New Haven and London: Yale University Press, 1985), ch. 2; Carl F. Kaestle, *Pillars of the Republic: Common Schools and American Society, 1780–1860* (New York: Hill & Wang, 1983), ch. 6; Madeline Grumet, *Bitter Milk: Women and Teaching* (Amherst: University of Massachusetts Press, 1988); Alison Prentice and Marjorie Theobald, "The Historiography of Women Teachers: A Retrospect," in *Women Who Taught: Perspectives on the History of Women and Teaching,* ed. Alison Prentice and Marjorie Theobald (Toronto: University of Toronto Press, 1991).

4. See Michael Apple, *Teachers and Texts: A Political Economy of Class and Gender Relations in Education* (New York: Routledge and Kegal Paul, 1986). Patrick Harrigan makes this argument in regard to Canada. See Harrigan, "The Development of a Corps of Public School Teachers in Canada, 1870–1980," *History of Education Quarterly* 32 (1992): 510.

5. The source for the percentages in 1837 and 1847 is the *Eleventh Annual Report of the Board of Education* (Boston, MA, 1848), 24. The source for the percentage in 1829 is Joel Perlmann and Robert A. Margo, *Women's Work?* American Schoolteachers, 1650–1920 (Chicago: University of Chicago Press, 2001), 28.

6. See "Table 1.2: Feminization of Teaching, 1829–60, Massachusetts and New York," in Perlmann and Margo, *Women's Work?* 28.

7. See David Murray, *Historical and Statistical Record of the University of the State of New York* (Albany, NY: Weed, Parsons & Co., 1885), 504. For a recent overview of the academy movement, see Kim Tolley and Nancy Beadie, "A School for Every Purpose: An Introduction to the History of Academies in the United States," in *Chartered Schools: Two Hundred Years of Independent Academies in the United States, 1727–1925,* ed. Nancy Beadie and Kim Tolley (New York: Routledge, 2002), 3–16.

8. *First Annual Report of the General Superintendent of Common Schools of the State of North Carolina* (Raleigh: W. W. Holden, 1854), NCC; Charles L. Coon, *The Beginnings of Public Education in North Carolina: A Documentary History, 1790–1840,* vol. 1 (Raleigh: Edwards & Boulton, 1908), NCC.

9. All of the teachers in the *Raleigh Register* also appear in the body of Charles L. Coon's documentary history (although not all appear in Coon's index).

10. Many newspaper advertisements named the teacher's place of origin, particularly when a teacher was a new arrival to the school. Nevertheless, these numbers must be interpreted with caution, because it is not possible to know with absolute certainty that a teacher whose origin is not specified came from New England or North Carolina. In her analysis of newspaper sources, Tolley presumed a teacher to be Southern if he or she was not specifically identified as "northern."

11. The figure of 121 schools is based on institutions appearing in the contents pages of Charles L. Coon, ed., *North Carolina Schools and Academies: A Documentary History.* Tolley's monthly interval sampling of the *Raleigh Register,* 1800–1840, NCC, did not reveal additional institutions of this sort. The proportion of schools known to have enrolled females was 50 percent in the fourth decade, but since it was not possible to determine the gender of the students from many newspaper sources, this figure may actually have been higher.

12. *First Annual Report of the General Superintendent of Common Schools of the State of North Carolina* (Raleigh, NC: W. W. Holden, 1854), NCC, 31.

13. See especially, Laurel Thatcher Ulrich, *The Age of Homespun: Objects and Stories in the Creation of an American Myth* (New York: Alfred A. Knopf, 2001) and *A Midwife's Tale: The Life of Martha Ballard, Based on Her Diary, 1785–1812* (New York: Vintage, 1991).

14. Margaret A. Nash, "'A Triumph of Reason': Female Education in Academies in the New Republic," in Beadie and Tolley, *Chartered Schools,* 64–88. The quotes are on pages 68 and 71, respectively.

15. Thomas Dublin, *Transforming Women's Work: New England Lives in the Industrial Revolution* (Ithaca, NY: Cornell University Press, 1994).

16. Lucy Larcom, quoted in Elisabeth Anthony Dexter, *Career Women of America, 1776–1840* (Clifton, NJ: Augustus M. Kelley, 1972), 1.

17. See Coon, *North Carolina Schools and Academies,* 390.

18. See Myra H. Strober and David Tyack, "Why do Women Teach and Men Manage? A Report on Research on Schools," *Signs: Journal of Women in Culture and Society* 5, no. 3 (1980): 494–503; Myra H. Strober and Audri Gordon Lanford, "The Feminization of Teaching: Cross-Sectional Analysis, 1850–1880," *Signs: Journal of Women in Culture and Society* 11, no. 2 (1986): 212–235. Also, see the summary of these and related arguments in Perlmann and Margo, *Women's Work?* 20–21, 27–28, 35–39, 101–106.

19. Lima's population numbered 1,764 in 1830. *Fifth Census of the United States, Bureau of the Census, Department of the Interior* (Washington, D. C.: Government Printing

Office). For further information on early settlement, community organization, and schooling in Lima, see Nancy Beadie, "Defining the Public: Congregation, Commerce and Social Economy in the Formation of the Educational System, 1790–1840" (PhD dissertation, Syracuse University, 1989).

20. *Record Book, Lima School District # 4, 1814–1854, Town Book, 1797–1818,* and *Town Book, 1818–1840,* LHS.

21. Nancy Beadie, "In the Pay of the Public: Changing Ideas about Gender and Political Economy in 19th Century New York" (paper presented at the Annual Meeting of the American Educational Research Association, San Diego, 2004).

22. Perlmann and Margo, *Women's Work?* 102.

23. Ibid., 35–39, 99.

24. See Kessler-Harris, *Out to Work: A History of Wage-Earning Women in the United States* (Oxford: Oxford University Press, 1982); Dublin, *Transforming Women's Work.*

25. See David B. Tyack, *Turning Points in American Educational History* (Waltham, MA: Blaisdell, 1967), 414; Kessler-Harris, *Out to Work,* 55; Perlmann and Margo, *Women's Work?* 32.

26. Kessler-Harris, *Out to Work,* 37.

27. "Report of the Managers and Treasurer, read at the Annual Meeting, July 29th, 1822," in *Revised Constitution and By-Laws of the Raleigh Female Benevolent Society, Adopted July 23rd, 1823* (Raleigh, NC: J. Gales & Son, 1828), http://socsouth.unc.edu/nc/benevolent/benevolent.html (retrieved April 11, 2003); *Raleigh Register,* June 8, 1827, in Coon, *The Beginnings of Public Education in North Carolina,* 209–210.

28. Dublin, *Transforming Women's Work.* The quotes are from pages 59 and 69, respectively.

29. There exists no systematic data on the wages paid to domestic servants prior to the 1850s. See Faye E. Dudden, *Serving Women: Household Service in Nineteenth-Century America* (Middletown, CT: Wesleyan University Press, 1983).

30. On the history of academies, see Beadie and Tolley, *Chartered Schools.* See also Fletcher Harper Swift, *A History of Public Permanent Common School Funds in the United States, 1795–1905* (New York: Henry Holt, 1911); Nancy Beadie, "Market-Based Policies of School Funding: Lessons from the History of the New York Academy System," *Educational Policy* 13, no. 2 (May 1999): 296–317; George Frederick Miller, *The Academy System of the State of New York* (Albany: J. B. Lyon, 1922).

31. Unfortunately, precise wage and board information for individual teachers exists for the academy only after 1832, while comparable information for the local district schools survives only for the period before 1834. Thus the data is directly comparable for only a single year, 1833.

32. Perlmann and Margo, *Women's Work?* 55 and 59. They cite a female/male wage ratio for the Mid-Atlantic region in 1832 of 0.42, and for New York State in 1860 of 0.41.

33. Data compiled from *Account Book #178, Journal of the Doings of the Legal Board of Trustees of the Genesee Wesleyan Seminary, 1830–1854,* GWSC. The index used to convert wages to constant value terms is the Composite Consumer Price Index from McCusker, *How Much is That,* 49–60.

34. Nancy Beadie, "Female Students and Denominational Affiliation: Sources of Success Among Nineteenth Century Academies," *American Journal of Education* 107, no. 2 (February 1999): 75–115.

35. Ibid. See also Beadie, "Analyzing the Impact of Female Student Markets on the History of Higher Learning in the United States," (paper presented at the Annual

Meeting of the History of Education Society, New Haven, Connecticut, 2001); also *Annual Reports of the Regents of the University of the State of New York* (Albany: State of New York, 1835–1880).

36. Beadie, "Market-Based Policies."

37. See newspaper advertisements in Coon, *North Carolina Schools and Academies,* 557–558, 616, and 605–606, respectively. The quote from Warrenton's Female Academy is on p. 606. Shady Grove Academy used the same language in its 1822 advertisement, on pp. 628–629.

38. *North Carolina Schools and Academies,* See advertisements on pages 103, 305, 559, and 629.

39. Ibid., 548.

40. See Tolley, "'A Comfortable Living for Herself and Her Children': The Gender and Wages of North Carolina Music Teachers in a Free Market, 1800–1840" (paper presented at the Annual Meeting of the Social Science History Association, Chicago, 2004). See also Tolley, *The Science Education of American Girls: A Historical Perspective* (New York: Routledge, 2002), chs. 4 and 7.

41. Beadie, "Market-Based Policies," 296–317.

42. In 1838, Regents academies derived 82 percent of their revenue from tuition, and 84 percent of the combined total revenues went to pay teacher salaries. See Beadie, "Market-Based Policies," 296–317.

43. Minutes of the Board of Trustees, Sherburne Academy, September 13, 1842, Sherburne Academy Collection, NYSHA.

44. Box #4, Minutes of the Board of Trustees, June 21st, 1849, Falley Seminary (aka Fulton Academy), Fulton Academy Collection, Fulton Public Library, Fulton, New York.

45. *Eleventh Annual Report,* 28.

46. Perlmann and Margo, *Women's Work?* 58.

47. McCusker, *How Much is That,* 110.

48. See Lance E. Davis, Richard A. Easterlin, and William N. Parker, ed., *American Economic Growth: An Economist's History of the United States* (New York: Harper & Row, 1972), 24–26; J. Leander Bishop, *A History of American Manufactures from 1608 to 1860* (Philadelphia: Edward Young, 1864), 179; See "Table A-1" in McCusker, *How Much is That,* 53–55.

49. Mark M. Smith, ed., *The Old South* (Malden, MA: Blackwell, 2001), xv–xvi, 2–3; Broadus Mitchell, *The Rise of Cotton Mills in the South* (Baltimore: Johns Hopkins University Press, 1921).

50. See the treasurer's reports in *Revised Constitution and By-Laws,* http://docsouth.unc.edu/nc/benevolent/benevolent.html (retrieved May 10, 2003).

51. Perlmann and Margo, *Women's Work?* 39–40.

52. Examples of sliding tuition scales in North Carolina academies and venture schools abound in Coon, *North Carolina Schools and Academies.* For example, see pp. 523–524 and 559, respectively.

53. See Kim Tolley, "A Chartered School in a Free Market: The Case of Raleigh Academy, 1801–1828," *Teachers College Record* 107 (January 2005): 59–88.

54. Kathryn Kish Sklar, "The Schooling of Girls and Changing Community Values in Massachusetts Towns, 1750–1800," *History of Education Quarterly* 33 (Winter 1993): 511–542.

55. See Ting-Hong Wong, "Education and State Formation Reconsidered: Chinese School Identity in Postwar Singapore," in this volume.

PART IV

Culture, Identity, and Schooling

9

From Spaniard to Mexican and Then American: Perspectives on the Southwestern Latino School Experience, 1800–1880

Victoria-María MacDonald and Mark Nilles

During the span of only half a century, residents of what became the southwestern states of the United States were transformed from subjects of the Spanish monarchy to citizens of the Mexican Republic (1821–1848) and then to conquered peoples of the United States as a result of the Mexican American War (1846–1848). As questions of citizenship and national identity shifted under these governments, the role and purpose of public education were transformed. In this chapter we explore the ways in which social categories and social relations were structured under these governments, and how the agency of Native Americans and Latinos has conditioned, challenged, and altered those categories and social relations in the realm of schooling. We particularly explore issues surrounding the political dynamics of language (indigenous, Spanish, English) and religion (non-Christianity, as practiced by local tribes, Catholicism, and Protestantism) as they related to the purposes of schooling. Under each form of governance (monarchy, republic, democracy), the role and purpose of public schooling held different meanings for the three major ethnic populations in the Southwest: Native Americans and their descendants; mestizos (the descendants of Native American and Spaniard unions); and Spaniards claiming pure blood lines to their European families. Of these three groups, mestizos were the largest and included a smaller subgroup of Afromestizos.

In this portrait of Latino schooling in the 1800s, we also intend to expand the traditional understanding of "colonial" and "early national" schooling in

U.S. educational historiography. The study of the development of education and schooling systems in the former colonies of England on the East Coast underplayed developments simultaneously occurring in other regions of the new United States, particularly the South and Southwest.[1] As historians of this era have expanded their focus to encompass broader intersections between cultures and countries in the Atlantic world, we also examine schooling within this more dynamic analysis of the nineteenth-century Americas.

Spanish Colonial Legacies

Formal and nonformal education in northern New Spain occurred within the context of Spanish exploration, conquest, and settlement. Conquistadores carried out these activities under the name of both the Crown and the church. As David Weber explained, the explorers believed they could "serve God, Country, and themselves at the same time."[2] The search for gold and other riches was no less a part of Spain's intent as it rose to international power in the fifteenth and sixteenth centuries. Although the quest for material wealth proved disappointing for the Spaniards who pushed into the American Southwest, it created a permanent imprint upon Native American culture. The Spanish imposed their language in verbal and written forms and brought with them the beginnings of formalized European education. The collision of cultures, languages, and religions over three centuries produced a new people who are the ancestors of present-day Southwestern Latinos.

The Spanish conquest had a devastating impact on Native American populations. Infectious European diseases (particularly smallpox), abuse of the *encomendero* (a system in which natives involuntarily worked Spaniard's lands), enslavement, and forced relocation into missions permanently altered the demographics of the Americas. Close contact between Europeans and Native Americans in the mission compounds often resulted in alarmingly high rates of mortality.[3]

Furthermore, colonial policies dictated the land, civil, and political rights of individuals in New Spain on the basis of their skin color, race, ethnicity, and national origin.[4] European-born *peninsulares* occupied the highest legal and social status in New Spain. The second tier consisted of criollos, individuals born in New Spain of pure Spanish parentage. Legally, Native Americans occupied the third tier. In return for accepting colonial rule, they were accorded land rights, and the church became their legal protector.[5] Socially, Indians were placed below both African slaves and free blacks.[6]

Adding to the complexities of the long centuries of settlement, many soldiers and settlers entered into unions (legitimate or illegitimate) with Indian women. The children of these unions were called mestizos. Mestizos occupied a nebulous legal and political status in the eyes of the Spanish Crown. Depending upon factors such as the status of the father and whether the parents were married in the church, some mestizos eventually "passed" into society as criollos.[7] In status-conscious New Spain these legal, social, and political distinctions impacted educational opportunities.

During the 1600s and 1700s, Spanish explorers pushed northward from central Mexico into the lands that form the contemporary United States. Both settlers and missionaries carried out formal and nonformal educational activities after the rudiments of living were established. Two formal types of education—settlers' schools and mission schools—emerged simultaneously, each reflecting the hierarchical nature of Spanish colonial society. Settlers' schools were charged with the preservation of Spanish language, culture, and religion in the New World. Run either by secular teachers or under the auspices of the missions, this formal education for cultural transfer was initially reserved only for the children of Spanish settlers, civil leaders, and military officers. Over the course of time, these schools also included children born in the New World, many of whom were mestizos.

The more prevalent type of formal education under Spanish rule occurred in the missions. Missions were generally enclosed compounds run by priests and lay brothers to Christianize and "civilize" the Native Americans. The educational function of the missions was reserved exclusively for Native Americans, as colonial policy dictated the separation of whites from Indians in the missions.[8] Unlike the social purpose of cultural *transfer* in settlers' schools, mission education was purposefully designed to *replace* Native American languages, religions, dress, and other cultural attributes with the Spanish language, Roman Catholic faith, and European mores and customs for current and future generations. Although some priests learned native languages to become more effective interpreters and agents of cultural transformation, the mission's role in the deculturalization of Native Americans was extensive.[9]

Settlers' Schools in the Colonies

The curriculum of the settlers' schools, a combination of classical learning and the Roman Catholic catechism, reflected the Spanish culture's close association of education with religion and its perception of education as a privilege for the elite. For example, the king's 1782 proclamation mandating Spanish-language schools stated as justification that they would offer "better instruction in the Christian doctrines and [improve] polite intercourse with all persons."[10] The schooling available to children of settlers, civil, and military leaders served the colonial era's rudimentary needs of literacy to communicate with Spanish officials, conduct trade, and in the case of priests, record baptisms, marriages, and deaths. Spanish settlers' schools thus represented a continuation and affirmation of their religious and linguistic heritage.

Mission Schooling and Culture

The most systematic and well-documented form of education available in northern New Spain occurred within the vast network of Catholic missions. It was within the missions that Native Americans experienced direct alteration of their culture, politics, economy, and demography. Once viewed by historians as paternalistic communities, the missions have been examined in recent years from the point of view of Native Americans using new techniques in ethnohistory, social history, and anthropology.[11] Native Americans who had spent time in the missions were often cast out of their former tribes and as a result remained in the mission. The missions' function of deculturalization powerfully shaped the Euro-Indian relationship. Native Americans attempted to preserve elements of their culture even within the constraints of highly circumscribed mission life.[12] However, the combination of military enforcement from soldiers armed with superior firearms and cavalry and the weakened state of tribes from disease and malnutrition placed Native Americans at a severe disadvantage to fight the Spanish conquest.

The missions' key role in forging and pacifying the Northern Frontier received official backing from the Crown. The state financed travel expenses, salary, and supplies for the priests and lay brothers. The missions were originally financed for ten years—the time considered sufficient to convert and "civilize" a Native American community. After that period Native Americans were supposed to be permitted self-governance and granted small land parcels. In practice, most missions lasted longer than ten years, particularly in the seventeenth and eighteenth centuries.

The educational goals of the missions were clearly articulated by the Crown. The actual practices that occurred were negotiated activities between Native Americans and missionaries. Missions played a key role in the frontier development of New Spain. As educational institutions, they were extremely organized, with highly educated teachers. The curriculum brought to Native Americans was a transformative force. Its role was to transfer and replace native languages with that of the colonizing Spanish power. For the first generation of children who learned to read and write Spanish (often called *ladinos*), their skills could become assets in a world that was permanently transformed economically, politically, and demographically. However, few opportunities for independent living were permitted prior to independence from Spain. In some cases, educated Indians fuelled rebellions and became leaders of warring tribes.[13] At the end of the colonial era, literacy skills were more practical when the missions were secularized and Native Americans were granted limited land rights and governance in the new pueblos. The intent of the missions, however, had never been to raise a generation of youth ready to challenge the status quo.

The Demise of Spanish Rule

During the late colonial era, Latin American-born individuals (criollos) grew increasingly dissatisfied with Spain's rigid and hierarchical policies. Specifically, Spanish-born citizens (peninsulares) were routinely placed in privileged church and state positions, marginalizing those born and educated in Latin America. Throughout the northern part of the South American continent, liberation from Spain was spearheaded by Simon Bolívar, called "the Liberator," or "the George Washington of South America." He not only freed Colombia, Peru, Ecuador, Venezuela, and Bolivia from Spain, but also spread ideas concerning citizenship and education throughout New Spain. Bolívar viewed universal public education as key to new republics and included compulsory education and the schooling of girls in their constitutions, both far-reaching reforms for the early 1800s.[14]

Within this political climate, Mexico's decade-long fight for its independence from Spain dates from 1810, when Catholic priest Miguel Hidalgo y Costilla delivered his famous "Grito de Dolores" (Cry of Dolores), demanding independence for the colonies.[15] In response to the revolutionary spirit brewing in the colonies, the liberal faction of the Spanish Cortes (parliament) inaugurated a new era in Spanish political history. The Cortes created a liberal constitution in 1812, requiring the king to become more responsive to the parliament; granting citizenship to all Spanish subjects, including Indians; and abolishing the Inquisition.[16] These radical reforms were threatened when Ferdinand VII regained the throne in 1814 and put a stop to these liberal measures. Eventually, the Spanish military revolted and restored the 1812 Constitution.[17] Despite these rapid political upheavals, the egalitarian principles of the constitution reached the Far Northern Mexican colonies in the early 1800s and, according to David Weber, its effects "continued to be felt long after independence."[18] The official independence of Spain from Mexico in 1821 inaugurated a tumultuous 25-year period of political, economic, and social change, with more rhetoric and words on paper than actual provision for schools.

Education during the Mexican Era, 1821–1848

The social, economic, and political changes accompanying the independence of Mexico from Spain in 1821 profoundly affected schooling in the Far Northern colonies. Most significantly, Mexican independence ended the official close relationship between education and religion that largely defined the colonial era. The end of state-sponsored religious missions, a new spirit of egalitarianism, and constitutional requirements for schooling combined to bring new importance to public schooling. Unlike before, public schools were identified as critical components in the creation of an educated citizenry. For the first time, citizenship training was articulated in public discourse and documents as a rationale for schooling. In this regard, the link between education and the republic echoed the Jeffersonian principles articulated in the early Republican Era of the United States.[19] However, decades of political upheaval in the fledgling Mexican state and the unintended negative consequences of the closure of the missions prevented the widespread establishment of public schools.

The Mexican Revolution altered a segmented colonial system that had subsidized university training for the elite and for the lowest tier (Native Americans), but provided little for the middle and working classes. Public education was thrust into debates over the shape and character of the new republic. Echoing the sentiments of many American republicans, the Mexican secretary of state argued, "Without education liberty cannot exist."[20] Mexico's 1824 Constitution adopted these ideals and required the provision of education for the masses. However, neither state nor federal funds were authorized for public schools, leaving local communities with the unrealistic burden of generating resources. Similar to many public schooling measures passed in the legislatures of the pre–Civil War American South that went unrealized, ambitious plans rarely resulted in permanent institutions outside of the largest and most prosperous communities of urban Mexico.[21] In general, many parallels exist between the evolution of Mexico's public school system during these formative years and efforts in the United States immediately prior to the common school movement of the 1830s and 1840s. Key reformers of American public schools—Horace Mann, Henry Barnard, and Catharine Beecher—had all claimed that locally run district schools and poor teacher preparation were the results of poorly organized and financially strained schools.[22] Similarly, the new Mexican nation aimed to remedy the consequences of colonial oversight.

Political, Economic, and Social Change in Mexico, 1821–1848

Many scholars date the origin of Mexican independence to the 1810 "Grito de Dolores," but it would be more than a decade before the criollo military officer Augustín de Iturbide officially declared Mexico independent from Spain in 1821.[23] The Mexican Revolution was chiefly an uprising of the Mexican-born creoles, or criollos, against Spanish domination, but egalitarian principles also stirred reformers during the tumultuous first few decades.[24] Historians Meyer and Sherman describe Mexican history during the 1830s and 1840s as one that "constantly teetered between simple chaos and unmitigated anarchy."[25] The average presidential term during the 1830s and 1840s was seven and one-half months. Because of delays in communication and transportation, government officials in the Far Northern frontier often had difficulty deciding which regime's policies and laws to follow. As a result, local officials often created their own laws or used outdated policies.[26]

The revolution's 1821 Plan de Iguala (equality) declared "the social and civil equality of Spaniards, Indians, and Mestizos."[27] The Mexican Era thus ushered in changes in social relations between the main populations of colonial society—Spanish, Indian, Black—and the many variations (mestizos, Afromestizos, etc.) that 300 hundred years of intermingling had produced. In the new nation, racial classifications in public documents were forbidden and replaced with the more general cultural term *gente de razón* (people of reason). Menchaca asserts that by the time of the Mexican Era, "gente de razón" was a term referring to Catholics and the racially mixed heterogeneous population that "practiced Spanish-Mexican traditions," but excluded tribal Indians.[28] Despite revolutionary ideals disparaging Spain's three-century-long racial and ethnic caste system, a racial hierarchy that prized white and devalued the darker-hued Indian, black, and Afromestizo members of its population continued during the Mexican Era (1821–1848).[29] Furthermore, slavery was not abolished, although liberal plans were enacted for gradual emancipation. In fact, the new constitution of 1824 banned slave trade, and required slave children to be set free at the age of 14, and adult slaves to be set free after 10 years of work.[30]

In addition to cultural changes within frontier Mexican society, economic shifts also impacted education. Historians generally agree that during the Mexican Era, trade and accumulation of capital increased stratification among social classes. The abolition of Spain's restrictive policies opened the door for trade with Americans, French, British, and Russians.[31] William Bucknell's opening of the Santa Fe Trail in 1821 facilitated overland trade between the East and West coasts. Ships from the American Northeast and Europe exchanged luxury goods and other manufactured products for raw products in the ports of California. Demand for cattle hides, tallow, and wool among manufacturers overseas contributed to the emergence of a wealthy Mexican ranchero class in Texas, California, and New Mexico that occupied the highest tier of Mexico's new social order.[32] This wealthy class was able to provide financial support to bolster local school initiatives.

The Spanish missions had been agricultural and educational settlements for three centuries; they, too, came under fire in the new order. In the eyes of many reformers, the 1821 Plan de Iguala and the 1824 Constitution declaring Native Americans as equal citizens to those of Spanish descent rendered the missions antiquated institutions of feudal Spanish society. In addition, many people felt that missionaries had monopolized valuable land for settlement, resulting in economic and political gain. These sentiments contributed to the government's decision to secularize the missions. Secularization ended the formal provision of education to Native Americans. Whether social, economic, or political, the cumulative effect of Mexican independence disrupted traditional hierarchies. Education was no longer reserved for only the elite members of society, nor was it a simply a tool to impose Christianization and Hispanicization upon indigenous peoples.

Decline of the Missions

During Spanish colonial rule missions were the primary cultural institutions on the frontier. These cultural, educational, and religious institutions became casualties of Mexican independence. The original role and purpose of the missions under Spanish colonial rule became anachronistic after independence. Several factors contributed to the end of the missions by the 1820s in Texas, Arizona, and New Mexico, and by the mid-1830s in California. Ideologically, the paternalistic and condescending attitudes of the Spanish priests toward Native Americans clashed with the liberal ideals of the independent government. In addition, the fight for independence created anti-Spanish and anticlerical feelings among the colonists. As a result, few Spanish-born priests could be persuaded to work on the frontier.[33] Finally, both the Spanish government during the rebellion and the Mexican government in the 1820s and 1830s directed funds away from the missions to military or other needs. Missionary priests were thus unable to adequately maintain their extensive holdings and continue to provide living expenses for Native Americans.[34]

Paths toward Mexican Public Schools

Under colonial rule, the purpose of public education was to transfer and maintain Spanish culture, its language, and the Catholic religion in the New World.

After 1821, the role and purpose of public schools shifted ideologically to reflect more egalitarian Enlightenment ideology and freedom from the yoke of monarchy. Growth in the actual number of public schools in the Mexican Era was less significant than the population's changing perception of the role of education in a republic. The historian Richard J. Altenbaugh found a similar pattern in U.S. history, arguing that schooling "underwent a profound transition, not so much institutionally but conceptually, during the early days of the [American] republic."[35]

The creation of public schools during the Mexican Era was uneven and depended upon the energies and resources of local communities. However, it was very characteristic of the "district school" stage of educational development in the U.S. frontier of the early 1800s. During the district school era, rural communities often obtained their own resources for schools through donations of money, firewood, and food. Furthermore, the qualifications of local teachers were often meager, and communities hired itinerant schoolmasters with few formal qualifications.[36]

The Mexican nation's antimonarchical republican spirit was reflected in both the rhetoric and administrative structure created to support education. Public education received formal recognition in the 1824 Constitution.[37] Unlike the U.S. Constitution, which does not mention education specifically, but declares in the Tenth Amendment that education is the responsibility of the states, the Mexican Constitution specifically mandated education in Article 50. Both primary and university schooling were mentioned in Article 50, which required the establishment of colleges for the "Marine, Artillery and Engineer Departments," and for teaching "the natural and exact sciences, the political and moral sciences, the useful arts and languages." Furthermore, the General Congress could "regulate the public education in their respective states," as long as Congress did not "prejudice" the "rights which the states possess."[38] Despite the broadminded and progressive attitude embodied in the constitution, the absence of a predetermined federal funding mechanism (e.g., taxes or land grants) weakened the implementation of the public school system.

The administration of Mexico's public education was heavily hierarchical. The Congress passed laws at the federal level that the various departments (states) were required to implement. At the departmental level, laws were then passed onto the local level. Power over local schools regarding finances, hiring and firing teachers, and other administrative matters rested with the local political body called the *ayuntamiento*. The ayuntamiento held considerable power over the fledgling public schools, like the mid-nineteenth-century U.S. school boards.

At least two major pieces of educational legislation were passed in the National Congress between 1821 and 1848. In 1833, the Congress issued sweeping and detailed legislation regulating public schools. Among the 19 articles in the 1833 School Act were those requiring the creation of normal schools and the appointment and pay of school inspectors to visit public schools. Furthermore, it was ordered that primary schools be opened at each of the six national colleges. Tensions between the Catholic Church and the state appear to underlie Article 9, which ordered fines for parishes or religious orders who were requested to open schools and "fail[ed] to do so." The inclusion of the Catholic Church as a provider of education also underscores a key difference between the Mexican and the U.S. traditions of education. Although the U.S. public schools stated separation between church and state, a strong Protestant ethic (and the initial use of the Protestant Bible) underpinned the public schools. The Mexican 1833 School Act further required annual public examinations at local schools and provisions for children who, "due to their poverty, deserve to be helped." Primary education for girls was also addressed. In addition to the prescribed curriculum of "reading, writing, arithmetic and the political and religious catechisms," it was stated that girls "shall be taught to sew, embroider, and other useful occupations of their sex."[39]

Three years later, in 1836, the Congress ordered departments to establish public schools in each pueblo. The ayuntamientos were also named as the chief administrative units, a function they fulfilled de facto for numerous years.[40] The legislative acts of the 1830s, designed to be a blueprint for the constitution's educational mandate, were implemented with varying degrees of success across the new Mexican nation. In large urban areas such as Mexico City, public primary and secondary schools were numerous, well supported, and in close proximity to institutions of higher education. In rural Mexico, most notably the Far Northern departments of California, New Mexico (which included Arizona), and Texas, the ability to find qualified teachers, adequate supplies, and even buildings presented a formidable challenge. Of course, this should come as no surprise since remote, rural schools in any nation are the hardest to establish and support.

Conclusions

The development of public schools in the Far Northern Mexico frontier began during 27 years of rapid political, economic, and social change. By 1844, Mexico as a whole had almost 60,000 students attending 1,310 public primary schools.[41] However, educational development was relatively slow in the frontier communities of New Mexico, Texas, and California. Most children were educated at home, at a neighbor's home, or in a private or public school. Among the elite, children were sent abroad to the United States, England, France, or even the Sandwich Islands.[42] The resources of the Mexican government between the 1820s and 1840s were directed toward maintaining inner political stability and protection from the warring Native Americans on the frontier. Tyler calculated that the Mexican government's expenditure on education was a mere fraction of its military expenditure during these tumultuous decades.[43] As a result, the constitutional requirements for schools and subsequent legislative acts remained only partially fulfilled during the Mexican Era. As Anglo settlers began moving into Texas in the 1820s and ultimately declared independence in 1837, Tejanos (Mexicans living in Texas) had established, at least in concept, the idea of government-sponsored schooling. Despite these changes, two aspects of schooling remained consistent in the transition from Spain to Mexican governance. The Spanish language as well as close, although unofficial, ties with the

Roman Catholic Church were maintained. The subordination of Mexican laws and customs to the U.S. conquerors subsequent to the Mexican American War was carried out at all levels—political, economical, and social—including in the schoolhouse.

Americanization and Resistance: Contested Terrain on the Southwest Frontier, 1848–1880

Beginning in the mid-nineteenth century, the United States began an era of expansionism, supported ideologically by the notion of "Manifest Destiny." Journalist John O'Sullivan, who argued that Providence had granted the United States a divine mandate to spread from coast to coast, coined this term in 1845. The ideology of the United States possessing a Manifest Destiny ultimately provided justification for the Mexican American War.

Since the early 1820s, increasing numbers of British, French, Americans, and Russians had begun settling in Mexico's frontiers. Mexico welcomed these settlers, particularly to Texas. The Mexican government offered inexpensive fertile land to the settlers. In exchange, the settlers were required to obey Mexican laws, learn Spanish, and convert to Catholicism. Furthermore, Mexico overlooked slave trade laws as an additional inducement to American slave owners.[44] The generous land distribution drew colonists, such as Stephen Austin, to Texas with hundreds of land-hungry families. The trickle into Texas became a flood, and by 1830 Americans outnumbered Mexicans in Texas by 25,000 to 4,000.[45] The new settlers largely ignored the unenforceable laws regarding Hispanicization and Catholic conversion.

The independence of Texas in 1836 as a sovereign republic and its subsequent U.S. annexation in 1845 paved the way for the United States to spread to the West Coast. A weak Mexican military and government, unresolved border disputes between the United States and Mexico, resurgent Indian threats, demand for western lands, and President James K. Polk's fear of a British or Russian invasion of California contributed to the U.S. decision to declare war on Mexico.[46]

The defeat of Mexico, ratified in the Treaty of Guadalupe Hidalgo (1848), altered the political, economic, and social lives of Mexicans. The adjustment from Spanish to Mexican rule was less abrupt than the American conquest. Spain and Mexico had at least shared Spanish, and Catholicism was the official religion of the state in both countries. At the time of the war, Mexicans were not immigrants to the region, but were colonized peoples on their own land. Articles VIII and IX of the Treaty of Guadalupe Hidalgo articulated the rights and responsibilities of 100,000 Mexicans who had been conquered. Under Article VIII, individuals had one year to become Mexican citizens or seek U.S. citizenship.

As the new territories entered statehood, their constitutions narrowed suffrage restrictions. According to the historian Martha Menchaca, the new constitutions only granted suffrage to Mexicans considered to be of the "white" race. Mestizos, Indians, African Americans, and Afromestizos were denied political rights. For example, in 1849, California granted the vote to "every white, male citizen of Mexico who shall have elected to become a citizen of the U.S."[47] Blacks and Indians were excluded from this citizenship, although they had previously been protected under the treaty. Furthermore, statutes barred "non-white" populations from practicing law, becoming naturalized citizens, and in many cases, marrying Anglos.[48] This racialization of Mexican peoples not considered "white" also extended to schooling. For instance, by the early 1860s, California's school code stipulated that "Indians" were among the nonwhite populations to be excluded from the regular public schools, and many Mexicans were part Indian.[49]

Both citizenship rights and land rights granted under the Treaty of Guadalupe Hidalgo became the cause of considerable conflict between newly arriving Anglos and native-born Mexicans. For instance, scholars of the Mexican experience have been highly critical of violations of the treaty that resulted in widespread land loss. As part of the treaty, Mexico ceded 500,000 square miles—including the contemporary states of California, New Mexico, Arizona, Utah, Nevada, and parts of Colorado and Wyoming—for 15 million dollars. Mexicans owned much of this land in large tracts. Articles VIII and IX of the treaty protected the rights of Mexicans to continue ownership of land. "In the said territories, property of every kind, now belonging to Mexicans not established there, shall be inviolably respected. The present owners, the heirs of these, and all Mexicans who may hereafter acquire said property by contract, shall enjoy with respect to it, guaranties equally ample as if the same belonged to citizens of the United States."[50] Under pressure from Anglo settlers wishing to take title to Western land, the U.S. Congress passed the Land Act of 1851. The act created boards of land commissioners in each new state and territory to adjudicate the validity of former Mexican land grant titles.[51] Between 1848 and 1900, Mexicans lost millions of acres in the Southwest. The historian Albert Camarillo identifies several factors leading to this tremendous loss of land, including long and costly legal battles before the Board of Land Commissioners, exploitation by lawyers and other unscrupulous Anglos, Mexicans' lack of English skills, "spendthrift practices" of the Californio elite, and land confiscation by squatters.[52]

The tangible aspects of the American conquest codified in citizenship and property law represented only some of the dramatic changes for Latinos in the nineteenth-century Southwest. Cultural conflict between the arriving Anglo-Protestant settlers and new Mexican-Americans was muted during the 1830s and 1840s but escalated during the Mexican American War and into the 1850s. Anglos arriving in Texas and California brought with them negative stereotypes of the character, religion, and racial composition of Mexicans. In general, Mexicans were disparaged as "greasers," immoral, sexually degenerate, indolent, "mongrels," "papists," and potentially subversive politicos.[53]

The belief in Anglo-Saxon Protestant superiority, which settlers brought to the Southwest in the mid-nineteenth century, resulted from a convergence of factors. Proponents of the Mexican American War viewed southwestern land as wasted in the hands of mongrel Mexicans. The Californian settler T. S. Farnham, for instance, declared in 1840 that Californios were an

"indolent, mixed race" and that "the old Saxon blood must stride the continent."[54] The racial mixing between Spaniards, mestizos, Native Americans, and
African Americans over three centuries particularly offended Americans grappling
with their own questions about race, in the context of African American slavery and fears of miscegenation. As a slave state, Texas attracted Southern white
migrants who viewed dark-skinned peoples with suspicion and suspected they
may have been tainted with African blood.[55] As these white supremacist views
were carried into the Southwest, they were often manifested as acts of violence,
racial slurs, and blatant discrimination, contributing to the continued social and
economic decline of former Mexican citizens.[56]

An additional factor shaping Anglo negativity toward Mexicans was their
Roman Catholic religion. Anti-Catholicism resurfaced in the mid-nineteenth
century United States as thousands of immigrants, largely Irish Catholics, arrived
in the 1830s and 1840s. On the East Coast, anti-Catholicism took the extreme
form of convent and church burnings. The formation of the anti-Catholic, anti-
immigrant political party—The Know-Nothings—marked the culmination of
anti-Catholic hysteria in the 1850s. According to party members, Catholics were
loyal only to the pope in Rome and thus represented a subversive threat.[57]

Mexicans, who had traditionally combined Catholic religious feast days with
municipal events, often protested the attacks on Catholicism.[58] Furthermore,
during the Mexican Era, public schools were often taught by priests or nuns or
in Catholic church buildings. The close alliance between the Catholic Church
and public schools disturbed the arriving Anglo-Protestants.

During the first decades after the American conquest, Mexicans resisted the
marginalization of their language, culture, and religion through several means.
Varying by locality and time, Mexicans were able to retain some of their rights,
intermarry, and assimilate with leading Anglo families as one strategy of survival. Others formed *mutualistas* (mutual aid societies) or participated in more
militant and extralegal organizations such as "Las Gorras Blancas" (the White
Caps). This famous resistance group, for example, sabotaged the introduction of
barbed wire fences in New Mexico ranching areas.[59]

Eventually, Anglos surpassed southwestern Latinos numerically, politically, and
economically. The process of becoming residents of a nation with a distinct linguistic, cultural, and religious heritage was often painful for many Latinos,
whose roots in the new U.S. lands stretched back to the late 1500s. Proponents
of public education, the U.S. society's primary vehicle for Americanization
among newly arrived immigrants to the East Coast, encountered unique
challenges in developing a secular educational institution within a historically
Spanish Catholic culture.

Education and Nation Building in Texas, 1836–1880

The Americanization of Texas Mexicans, or Tejanos, began with the Texas
Revolution of 1836.[60] The Anglo-led government of the Republic of Texas
valued the potential benefits of a public school system. Even though Mexico
had created public schools in its constitution and in Texan land between 1821

and 1848, the Republic of Texas cited a lack of schools as one of its rationales for rebellion, stating, "It [the Mexican government] has failed to establish any public system of education, although possessed of almost boundless resources, (the public domain,) and although it is an axiom in political science, that unless a people are educated and enlightened, it is idle to expect the continuance of civil liberty, or the capacity for self-government."[61] Resolutions for public education were passed in the Congress of Texas, particularly the provision of land grants for schools, and newspaper editorials bemoaned the lack of genuine interest in supporting schools. However, citizens of the Republic of Texas encountered the same difficulties characteristic of Mexican Texas. Similar to most frontier areas, a combination of private, religious, and quasi-public institutions arose where enough students and a qualified teacher could be procured.[62]

In 1845, Texan citizens voted to be annexed by the United States, a step the latter was eager to approve. As residents of a newly acquired U.S. territory, Tejanos began to see their traditions, culture, and language come under fire. The establishment of American-style public schools after statehood in 1845 brought the Tejanos into further conflict with the rapidly growing numbers of Anglo settlers.[63] In the decades following the Texas government's 1854 "Act to Establish a System of Common Schools," Tejanos wishing for public education experienced shifting attitudes toward Spanish language use, equal access, and the employment of Tejanos or other Latinos as teachers.

In the urban centers of the Northeastern United States, public schools had long served the function of assimilating immigrants.[64] The southwestern experience differed because of preestablished Hispanic communities and the strong influence of German immigrants. Furthermore, Texas was admitted into the Union as a slave state, and after the end of the Civil War, it established public schools that legally segregated black and white children into different schools. Through law, the Texas Anglo-dominated legislature reinforced English as the public schools' proper language of instruction. Two years after Texas formally established public schools, an 1856 amendment stipulated that "no school shall be entitled to the [monetary] benefits of this act unless the English language is principally taught therein."[65] The amendment was approved again in Chapter 98, Section 9, in the 1858 legislature.[66] The state requirement of English-language instruction in Texas' nineteenth-century public schools was rarely fully adhered to; rural communities especially strayed from the requirement. Both German immigrants and Tejanos maintained their native languages in many public schools during the transitional decades of the 1850s through the 1880s.

Tejanos did not so much reject the English language as want to preserve Spanish while also learning the language of their conquerors. Thus, for example, the Spanish-language newspaper in San Antonio, *El Bexareño,* advocated that public education be conducted in both languages.[67] Shifting state policies reflected the fluidity of Americanization measures. When the public school system in Texas was re-created in 1871 under Radical Rule during Reconstruction, a more flexible approach was adopted toward the language interests of Tejanos, Germans, and French. In his first annual report, Superintendent of Public Instruction J. C. DeGress stated that as a result of "the large proportion of citizens of German and Spanish birth and descent in our State," teachers would be permitted to teach the German, French, and Spanish languages, "provided the time so occupied should not exceed two hours each day."[68] Public school officials in other locales, such as among the German communities of the Midwest, had also compromised on language policies in order to keep immigrant children in the public school system.[69]

The presence of the Spanish language and the use of Tejano teachers persisted during these transitional decades of Texan public schools. The enormous size of Texas and its rural nature contributed to a variety of local arrangements, some of which blended the Tejano community members with the schools. For example, on large ranches, owners created special schools for vaqueros' (cowboys') children. Students at the Randado Ranch in Zapata County and Los Ojuelos in Encinal County studied English and Spanish and took exams in three subjects at the end of each year.[70] The lines between public and private schooling were often blurred, particularly under the state's "community system." In the late nineteenth century, rural areas were permitted to use public funds for a school if the teacher could pass the Board of Examiner's test. Thus, for instance, the sisters of Nazareth Academy became certified by the state of Texas as community teachers and taught Mexican pupils in Spanish and English with public funds.[71]

Tejano parents seeking educational opportunities for their children also relied significantly upon the rapidly growing parochial schools sponsored by the Roman Catholic Church. According to San Miguel, Jr., Catholic schools were popular among Latinos in the Southwest for three reasons. First, Catholic schooling was seen as a form of preserving Latino identity because of the closely intertwined nature of religion and Latino culture. Second, the Catholic Church was willing to permit the speaking of Spanish in school and allow Mexican Americans to preserve other cultural traditions. Thus, instead of imposing "subtractive" measures upon Mexican children, measures San Miguel, Jr. and Valencia define as ones that not only "inculcate American ways, but also . . . discourage the maintenance of immigrant and minority group cultures," Catholic teachers permitted "additive" measures such as bilingual or trilingual language instruction.[72] Lastly, the recruitment of mostly female teaching orders provided an inexpensive method of staffing schools, and some of the sisters were native Spanish speakers from Spain or Mexico.[73] Both male and female teaching orders were also heavily recruited from France to the Southwest during this era, and they brought a liberal attitude toward the value of learning and teaching several languages.

As Texas changed its status from an independent republic to part of the United States, Tejanos experienced a political, social, and economic shift between 1848 and 1900. They attempted to seek the best educational alternatives for their children that local circumstances would allow. In some cases, parents viewed Americanization as the best strategy for adaptation and sent their children to the new public schools. Others selected Protestant schools, which may not have charged fees, such as the Presbyterian Day School in 1840s San Antonio. Similarly, Anglo-Protestants enrolled their children in Catholic parochial schools, which appeared to offer a superior education compared to the limited public school alternatives. During the latter half of the nineteenth century, the Texas public school system began to impose harsher restrictions against the Spanish language and created separate Mexican schools. In response, some middle-class Tejano communities established their own private schools, such as El Colegio Altamirano in Jim Hogg County (1897–1958), which were free from the control of the Catholic Church or the state of Texas.[74]

From Alta California to the State of California: Education, Americanization, and the Californio Population

> From a want of any organized system of school instruction while California remained a Mexican province, it is not surprising that, in very many cases, the children of the older Californians have little or no education beyond that of repeating and a few reading the ceremonies and religius [sic] books of the Catholic church. It is true that there are exceptions to the position taken; but scarcely in sufficient numbers to form any considerable amount. This class of our population has heretofore been deprived of the advantages of schools; and now, since the parents of such children have been brought into contact with the Anglo Saxon race, the want of education becomes more apparent to them, and they are alive to the interests of this important subject.
>
> John G. Marvin, Superintendent of
> Public Instruction of the State of Texas, 1851[75]

As in to Texas, a small number of Mexicans controlled a considerable portion of the land and political power in California at the time of the American conquest in 1848. Historians estimate that these elites represented only about 5 percent of the Mexican population. In order to distance themselves from the negative connotations ascribed to Mexicans, they began using the name Californio.[76] However, even those of the elite Californio class found their Hispanic language, culture, and religion under assault by the State of California's new constitution and laws.

After the U.S. conquest, the Gold Rush of 1849 brought thousands of Anglo settlers to the Golden State. By 1860, the Anglo and Mexican populations of California exceeded 380,000.[77] During this era of rapid change, the public school system of California was viewed as a stable influence in molding the diverse groups of European immigrants, Californios, and Anglos into the future citizens of California. Unlike Texas, which experienced a long disruption in the growth of its public schools owing to the Civil War and Reconstruction, California moved forward quickly in its public school development. A state superintendent was appointed in 1851, and schools were created in counties throughout the state. During the 1850s and 1860s, Californios and Anglos clashed over language issues in the new public schools. Early public school reports illuminate the bilingual/bicultural environment present in many communities. Like the Tejanos, Californios did not reject the English language, but also wished to preserve Spanish in both the home and the public domain.

The experience of Santa Barbara's Californios in the 1850s illuminates tensions over language issues throughout the Southwest, as Mexicans found themselves becoming foreigners in their native lands. Because Californios comprised three-fifths of the population, they initially wielded considerable influence in the community. Spanish language instruction was maintained in Santa Barbara's public schools in the early 1850s. Two male teachers from Chile taught geography, history, writing, and arithmetic.[78] However, the state of California began passing "subtractive" policies in 1855 that forbade the teaching of Spanish in public schools. The city's two Anglo school commissioners (the third was Mexican) called for English-only public schools. A temporary compromise was reached with the creation of a separate English school. However, the expense of maintaining two schools was prohibitive, and the English- and Spanish-speaking public schools were combined into one bilingual school. Upset, Anglo parents withdrew their children from the bilingual school. The city's increasingly anti-Mexican newspaper, the *Gazette*, declared "the parents of American children unwilling that they should learn a confused jargon and gibberish, prefer to keep them at home." By 1858 the Anglo parents had won the battle for English-only instruction. Subsequently, many Mexican parents chose to enroll their children in Catholic schools that permitted Spanish.[79]

The racialization of Latino children after the American conquest further shaped their educational circumstances in the post–Treaty of Guadalupe Hidalgo decades.[80] Most Mexicans in California and the Southwest were of varying degrees of *mestizaje*—the result of Native Americans, Spaniards, Africans, and Mexicans uniting over several centuries. Californios of darker complexion were placed at a distinct disadvantage if they classified themselves or were classified by the government as Native Americans (or blacks) instead of Mexicans. Menchaca suggests that as a result, many mestizos identified themselves to the government as Mexicans to avoid discrimination. For example, as early as 1858, "Negroes, Mongolians, and Indians" in California were not allowed to attend schools for white children "under penalty of the forfeiture of the public school money by districts admitting such children into school."[81] The revised California School Law of 1866 permitted some exceptions to this rule. Section 56 permitted a board of trustees to "admit into any public school half-breed Indian children, and Indian children who live in white families or under guardianship of white persons by a majority vote." Section 57 held, "Children of African or Mongolian descent, and Indian children not living under the care of white persons, shall not be admitted into public schools." School boards were required, however, to open public schools whenever at least ten parents of "such children" formally petitioned for schooling.[82] Except for the very few children who may have had white guardians willing to petition for entry to white schools, the majority of mixed race children were placed in segregated schools. The description of one such school in late nineteenth-century Los Angeles underscores this point:

> There is also a small school of fifteen negro children of all the shades arising from blending all the primary colors of Spanish, American, Indian, and African parentage. They are engaged in the pursuit of knowledge under difficulties, as their little room ten by fifteen feet, has neither desks, blackboard, maps, charts, nor any kind of furniture, except a line of rough board seats without backs, around the walls.[83]

Anglo attitudes regarding the educability of Mexicans, Indians, and African Americans were commonly disparaging. John Swett, superintendent of public instruction for California from 1863 to 1868, often made comments such as "the boys' school, numbering, say forty scholars, held in a comfortable brick school-house, is attended mostly by children whose mother tongue is Spanish, and who are not remarkable either for order or scholarship."[84] In San Buenaventura, Swett drew the conclusion that since "the American residents there have established a private school and refuse to send their children to the public schools, where the 'native' children attend, we are led to suppose that its management is not the best in the world."[85] The decades of cultural conflict between Californios and Anglos resulted in Mexicans' and their descendants' diminished political and economic status. However, as historians Richard Griswold del Castillo and Albert Camarillo have found, Mexican communities increasingly relied upon themselves to preserve their language, culture, and identity.[86]

Hispano Schooling in the Territories of New Mexico, Colorado, and Arizona

Both demographics and geography shaped an educational history in the territories of Arizona (1848–1912), Colorado (1848–1876), and New Mexico (1848–1912) that was distinct from that of California and Texas. The rural and isolated nature of the land and majority Hispanic citizenry preserved the Spanish language and led to blended public-private schools for a longer duration than in California and Texas. Although Arizona, Colorado, and New Mexico each present unique case studies, we will concentrate on the case of New Mexico. Again, two cultural components—religion and language—provide much of the context to the history of Hispano schooling.

Despite the ability of Hispanos to control their schooling options to a greater degree in New Mexico, tensions between the Anglos and Hispanos existed. One key example of the tensions between New Mexicans, or Hispanos, and Anglo settlers concerned the scope of Catholic involvement in public schooling and the centralized (often Anglo) territorial control of schools versus the local control preferred by Hispanos.

Established as a territory in 1850, New Mexico passed its first school law in the 1855–1856 legislature. The law stipulated that the schools would be supported by a property tax and that control over the schools would reside with the territorial government. The new law was unpopular with Hispanos, who were accustomed to local control and holding large tracts of land. Quickly, Hispanos repealed the law by a near-unanimous 99.3 percent majority against (5,016 to 37). Anglos viewed the repeal as a rejection of education in general, instead of a rejection of less control and more taxes. The territorial governor, William Pile, condemned the vote saying, "If more proofs of the present unfortunate condition of the mestizos were wanting, it may be shown that their indifference to education reaches not only hostility, but a hostility which has, perhaps, been expressed with more unanimity at the ballot-box than any similar instance in history." As Lynn Marie Getz documents in her skillful study of

Hispanos and education, the 1856 vote represented a "myth of Hispano resistance" to public schooling. In subsequent legislation, Hispanos supported public schools and even the idea of taxing themselves—but they wanted to maintain control over local funding.[87]

The Hispano community dominated New Mexico's school leadership at both country and territorial levels and subsequently possessed the type of political power necessary to protect its interests and concerns. The Hispano dominance is evidenced not only by the laws and policies written during the late nineteenth century, but also by the fact that the laws were published in both English and Spanish.

Of all the southwestern states, New Mexico appeared most able to embrace Spanish/English bilingualism in its public schools. In 1875, for example, two-thirds of the public schools were conducted exclusively in Spanish, and one-third were taught in Spanish and English. Only 5 percent of the territory's public schools were taught exclusively in English. Section 1110 of the 1889 School Law permitted texts in either language, stating, "It shall be the duty of the School Directors to adopt text books in either English or Spanish, or both."[88] Thus, the English-only rules, which characterized Texas and California by the 1850s, did not take root in New Mexico. By comparison, the territory did not require by law that English be included in all of the public schools until 1891.

The Catholic Church played an extremely influential role in the educational development of Latinos particularly and New Mexicans in general. A significant irritant to Anglo officials in the territory was the granting of public funds to Catholic schools. In his 1875 annual report, the territory secretary and ex officio superintendent of education William G. Ritch condemned the fact that "in a majority of the counties, today, the school books and church Catechism, published by the Jesuits, and generally in Spanish, constitute the text books in use in the public schools."[89] The separation of church and state that Ritch desired did not materialize in New Mexico during the nineteenth or even early twentieth centuries. County school commissioners often requested religious orders to organize public schools and paid priests and nuns from the common school fund.

Statistics from the territorial secretaries' reports suggest that Hispanos heavily patronized the Catholic schools and also opened their schools to non-Catholics. However, religious schooling in New Mexico was not limited to Catholic-centered education. Protestant missionaries, including the Presbyterian and Methodist Churches, viewed New Mexico as a missionary field ripe with possibilities. The schools were determined to save Latino children from Catholicism and "ignorance" in general.[90] These Protestant missionary schools created dilemmas for Hispano parents in New Mexico. The large number of children enrolled suggests that depending upon the availability of other schools, a missionary school was better than no school at all. However, many Hispano families were threatened with excommunication from the Catholic Church for sending their children to Protestant schools.

At the same time, historians point out that there could be long-term benefits for the children attending missionary institutions. Because most public schools were conducted in Spanish, and Catholic schools in Spanish and English, missionary schools "offered upwardly mobile Hispano entry to the language, values, and milieu of the Anglo world."[91] Specifically, Susan Yohn found that Hispano youth, especially those who converted, became "recipients of whatever largesse the mission enterprise had to reward" and became well connected with networks of Anglos who could further their education and assist with employment.[92]

Conclusions

Under the Spanish government, schooling for the privileged and for the conquered reflected the segmented priorities of an imperial power as it asserted its military, economic, political, and cultural dominance upon a new hemisphere. Home control of the colonies in the wave of independence begun by Simon Bolívar incorporated the ideals of the French and American revolutions. This Enlightenment ideology included the creation of public schools to facilitate republican participation in government. To accomplish this end, the spread of literacy was government sanctioned (although not always fully funded) during Mexico's nation-building era. Insufficient time, political disruptions, and lack of funds halted the successful implementation of Mexico's sweeping public school measures enacted in the 1820s and 1830s.

U.S. annexation and conquest of the Southwest occurred during the heyday of antebellum school reform. Consequently, former Mexican citizens encountered public schools bent upon rapid assimilation and Americanization into the English language and a pan-Protestant public school curriculum. Roman Catholic and private schools offered alternatives to the often-subtractive culture of many public schoolrooms. Region and social class strongly influenced the pace of Americanization and acceptance or rejection of the Spanish language and culture in the public schools of the late nineteenth century. Complex hybrids of Spanish conquest, Mexican governance, and U.S. annexation, schools in the Southwest were fluid entities, shaped both by those who inhabited their rooms, and by the political actors who created them.

Notes

1. Victoria-María MacDonald, "Hispanic, Latino, Chicano, or 'Other'? Deconstructing the Relationship between Historians and Hispanic-American Educational History," *History of Education Quarterly* 41(Fall 2001): 365–413.
2. David J. Weber, *The Spanish Frontier in Northern New Spain* (New Haven and London: Yale University Press, 1992), 23.
3. David Sweet, "The Ibero-American Frontier Mission in Native American History," in *The New Latin American Mission History*, ed. Erick Langer and Robert H. Jackson (Lincoln and London: University of Nebraska Press, 1995), 1–48.
4. Martha Menchaca, *Recovering History Constructing Race: The Indian, Black, and White Roots of Mexican Americans* (Austin: University of Texas Press, 2001).
5. Ibid., 81.
6. Lisbeth Haas, *Conquests and Historical Identities in California, 1769–1936* (Berkeley: University of California Press, 1995), 29.
7. Magnus Mörner, *Race Mixture in the History of Latin America* (Boston: Little, Brown and Company, 1967).

8. Rev. J. A. Burns, *The Catholic School System in the United States: Its Principles, Origin, and Establishment* (New York: Benziger Brothers, 1908), 507.

9. Martha Menchaca, "The Treaty of Guadalupe Hidalgo and the Racialization of the Mexican Population," in *The Elusive Quest for Equality: 150 Years of Chicano/Chicana Education,* ed. José F. Moreno (Cambridge, MA: Harvard Educational Review, 1999), 3–29.

10. Frederick E. Eby, comp., *Education in Texas: Source Materials,* University of Texas Bulletin, no. 1824 (April 1918) 6.

11. Robert H. Jackson, "Introduction," in Langer and Jackson, *Latin American Mission History,* vii–xviii.

12. Sweet, "Ibero-American Frontier Mission," 1–48.

13. See example of the youth named Frasquillo who, after being trained to read and write Spanish and Latin, turned on his mentor in the 1680 revolt. "When the conspiracy was formed and the day for the massacre was fixed, this precocious boy entered ardently into it." Burns, *Catholic School System,* 209–210.

14. Luis B. Prieto, *Simon Bolívar: Educator,* trans. James D. Parsons (New York: Doubleday), 1970.

15. David J. Weber, *The Mexican Frontier, 1821–1846: The American Southwest under Mexico* (Albuquerque: University of New Mexico Press, 1982); and Manuel G. Gonzales, *Mexicanos: A History of Mexicans in the United States* (Bloomington and Indianapolis: Indiana University Press, 1999).

16. Weber, *Mexican Frontier,* 16–17.

17. The evolution of Mexico's federalist monarchy and then federal republic during these decades is complex and detailed in works such as Jaime E. Rodríguez O., "The Constitution of 1824 and the Formation of the Mexican State," in *The Origins of Mexican National Politics, 1808–1847,* ed. James E. Rodriguez O. (Wilmington, DE: Scholarly Resources, 1997), 65–84; and Timothy E. Anna, "Augustín de Iturbide and the Process of Consensus, " in *The Birth of Modern Mexico, 1780–1824,* Christon I. Archer (Wilmington, DE: Scholarly Resources, 1997), 187–204.

18. Weber, *Mexican Frontier,* 17.

19. Gordon Lee, ed., *Crusade against Ignorance: Thomas Jefferson on Education* (New York: Teachers College Press, 1961).

20. Report of the Mexican Secretary of State, November 1823, quoted in *A Citizen of the United States, A View of South America and Mexico, Comprising Their History, the Political Condition, Geography, Agriculture, Commerce, etc. of the Republics of Mexico, Guatemala, Colombia, Peru, the United Provinces of South America and Chili, with a Complete History of the Revolution, in Each of These Independent States,* 2 vols in one (New York: Published for Subscribers, 1827), 125.

21. David N. Plank and Rick Ginsberg, eds., *Southern Cities, Southern Schools: Public Education in the Urban South* (New York: Greenwood Press, 1990); Victoria-Maria MacDonald and Mark R. Nilles, "The Persistence of Segregation: Race, Class, and Public Education in Columbus, Georgia, 1828–1998" (unpublished book manuscript, 2003); Edgar Knight, ed., *A Documentary History of Education in the South before 1860,* vol. 2 (Chapel Hill: University of North Carolina Press, 1949–1953.)

22. Carl F. Kaestle, *Pillars of the Republic: Common Schools and American Society, 1780–1860* (New York: Hill and Wang, 1983).

23. Gonzales, *Mexicanos.*

24. Michael C. Meyer and William L. Sherman, *The Course of Mexican History* (New York: Oxford University Press, 1995).

25. Ibid., 324.

26. Weber, *Mexican Frontier.*
27. Ibid., 47.
28. Menchaca, *Recovering History,* 167.
29. Ibid., 161–186.
30. Ibid., 163.
31. Weber, *Mexican Frontier,* 146.
32. Gonzales, *Mexicanos,* 65; Weber, *Mexican Frontier,* 208–210.
33. Matt S. Meier and Feliciano Ribera, *Mexican Americans—American Mexicans: From Conquistadores to Chicanos* (New York: Hill & Wang, 1993), 32–33; and Weber, *Mexican Frontier,* 44.
34. Weber, *Mexican Frontier,* 44.
35. Richard J. Altenbaugh, *The American People and Their Education: A Social History* (Upper Saddle River, NJ: Pearson Education, 2003), 45.
36. Kaestle, *Pillars of the Republic;* Lawrence A. Cremin, *American Education: The National Experience, 1783–1876,* (New York: Harper & Row, 1980); David B. Tyack, *Turning Points in American Educational History* (University Microfilms International, 1967).
37. Report of the Mexican Secretary of State, November 1823.
38. Eby, *Education in Texas,* 27.
39. Ibid., 85–88.
40. Daniel Tyler, "The Mexican Teacher," *Red River Valley Historical Review* 1(1974): 207–221.
41. Ibid., 209.
42. For information on children sent abroad see Tyler, "Mexican Teacher," 211; Irving B. Richman, *California under Spain and Mexico 1535–1847* (New York: Houghton Mifflin, 1911), 345–346.
43. Tyler, "Mexican Teacher," 209.
44. Weber, *Mexican Frontier,* 213.
45. Gonzales, *Mexicanos,* 70.
46. Historians have offered multiple interpretations of the motives of the Mexican American War. See, for example, John D. Eisenhower, *So Far from God: The War with Mexico, 1846–1848* (New York: Random House, 1989).
47. Richard Griswold del Castillo, *The Treaty of Guadalupe Hidalgo: A Legacy of Conflict* (Norman and London: University of Oklahoma Press, 1990), 66.
48. Menchaca, *Recovering History.*
49. Charles Wollenberg, *All Deliberate Speed: Segregation and Exclusion in California Schools, 1855–1975* (Berkeley: University of California Press, 1978), 13.
50. Del Castillo, *Treaty of Guadalupe Hidalgo,* 190.
51. Ibid., 73.
52. Albert Camarillo, *Chicanos in a Changing Society: From Mexican Pueblos to American Barrios in Santa Barbara and Southern California, 1848–1930* (Cambridge, MA and London: Harvard University Press, 1979), 114.
53. Arnoldo De León, *They Called Them Greasers: Anglo Attitudes toward Mexicans in Texas, 1821–1900* (Austin: University of Texas Press, 1983); Reginald Horsman, *Race and Manifest Destiny: The Origins of American Racial Anglo-Saxonism* (Cambridge, MA and London: Harvard University Press, 1981), 208–248.
54. Horsman, *Race and Manifest Destiny,* 210.
55. De León, *They Called Them Greasers,* 14–23.
56. Menchaca, *Recovering History;* Camarillo, *Chicanos.*
57. Jay P. Dolan, *The American Catholic Experience: A History from Colonial Times to the Present* (Garden City, NY: Image Books, 1987), 202, 295.

58. See Document 16, "Response to Know-Nothing Attacks, San Antonio, 1855," in *¡Presente!: U.S. Latino Catholics from Colonial Origins to the Present,* ed. Timothy Matorina and Gerald E. Poyo (Maryknoll, NY: Orbis Books, 2000).

59. Robert J. Rosenbaum, *Mexicano Resistance in the Southwest: "The Sacred Right of Self-Preservation"* (Austin and London: University of Texas Press, 1981); Deena J. González, *Refusing the Favor: The Spanish-Mexican Women of Santa Fe, 1820–1880,* (New York and Oxford: Oxford University Press, 1999).

60. *Tejano* was a term Hispanics in Texas utilized to identify themselves as early as 1833. See "TEJANO," *The Handbook of Texas Online,* http://www.tsha.utexas.edu/handbook/online/articles/view/TT/pft7.html (accessed August 15, 2003).

61. Eby, *Education in Texas,* 130.

62. See Eby, *Education in Texas,* 130–199, for information on Texas during the Republic.

63. Arnoldo De León, *The Tejano Community, 1836–1900* (Albuquerque: University of New Mexico Press, 1982), 1–22.

64. David K. Cohen, "Immigrants and the Schools," *Review of Educational Research,* 40 (February 1970): 13–28.

65. Eby, *Education in Texas,* 336.

66. H. P. N. Gammel, compiler and arranger, *The Laws of Texas, 1822–1897,* 10 vols (Austin, TX: Gammel Book Co., 1898), 998–999.

67. De León, *Tejano Community,* 197.

68. *First Annual Report of the Superintendent of Public Instruction of the State of Texas, 1871* (Austin: J.G. Tracy, State Printer, 1871), 10. Wisconsin State Historical Society.

69. Perlmann, "Historical Legacies," 27–37.

70. De León, *Tejano Community,* 191.

71. Carlos E. Castañeda, *Our Catholic Heritage in Texas, 1519–1936,* vol. 7, *The Church in Texas since Independence, 1836–1950.* (Austin: Von Boeckmann-Jones, 1958), 307.

72. Guadalupe San Miguel, Jr., and Richard R. Valencia, "From the Treaty of Guadalupe Hidalgo to Hopwood: The Educational Plight and Struggle of Mexican Americans in the Southwest," *Harvard Educational Review* 68, no. 3 (Fall 1998): 358.

73. Guadalupe San Miguel, Jr., "The Schooling of Mexicanos in the Southwest, 1848–1891," in *The Elusive Quest for Equality: 150 Years of Chicano/Chicana Education,* ed. José F. Moreno (Cambridge, MA: Harvard Educational Review, 1999), 31–51.

74. Cinthia Salinas, "El Colegio Altamirano (1897–1958): The Educational Opportunities of Tejanos in Jim Hogg County" (paper presented at the American Educational Research Association, Seattle, WA, April 2001); Guadalupe San Miguel, Jr., *"Let All of Them Take Heed: Mexican Americans and the Campaign for Education Equality in Texas, 1910–1981* (Austin: University of Texas Press, 1988), 10.

75. *Report of the Superintendent of Public Instruction of the State of California* (Sacramento, CA: J. B. Devoe, State Printer, April 10, 1851). Document in the Rare Book Collection, Library of Congress.

76. Leonard Pitt, *The Decline of the Californios: A Social History of the Spanish-Speaking Californians 1846–1890* (Berkeley: University of California Press, 1966), 7.

77. Menchaca, *Recovering History* 265.

78. *First Annual Report of the Superintendent of Public Instruction, California, 1852.* (Sacramento, CA: Eugene Cassely, State Printer, 1872), 49. Wisconsin State Historical Society (WSHS).

79. Camarillo, *Chicanos,* 17.

80. Menchaca, *Recovering History,* 256–267.

81. "Historical Sketch of the Public School System in California," in *First Biennial Report of the Superintendent of Public Instruction of the State of California for the School Years 1864 and 1865,* Special Collections, Monroe C. Gutman Library, Harvard Graduate School of Education, Harvard University, Boston, 258.

82. *Revised School Law, State of California,* Approved March 24, 1866 (Sacramento, CA: O. J. Clayes, State Printer, 1866), 18.

83. "John Swett on Schools in Southern California (1865)," in *Education in the United States: A Documentary History,* 2 vols. comp. Sol Cohen (New York: Random House, 1965), 1034.

84. Ibid., 1038.

85. Ibid., 1039.

86. Camarillo, *Chicanos;* Del Castillo, *Treaty of Guadalupe Hidalgo.*

87. Lynn Marie Getz, *Schools of Their Own: The Education of Hispanos in New Mexico, 1850–1940* (Albuquerque: University of New Mexico Press, 1997), 13–28.

88. *Compilation of the School Laws of New Mexico/Leyes de Escuelas de Nuevo Mexico* (East Las Vegas: J. A. Carruth, Printer, 1889), 14.

89. Ibid.

90. Examples of works on this topic include Susan M. Yohn, *A Contest of Faiths: Missionary Women and Pluralism in the American Southwest* (Ithaca, NY: Cornell University Press, 1995); Mark Banker, *Presbyterian Missions and Cultural Interaction in the Far Southwest, 1850–1950* (Urbana: University of Illinois Press, 1993); Jerry A. David, "Matilda Allison on the Anglo-Hispanic Frontier: Presbysterian Schooling in New Mexico, 1880–1910," *American Presbyterians* 74 (Fall 1996): 171–182; Norman J. Bender, *Winning the West for Christ: Sheldon Jackson and Presbyterianism on the Rocky Mountain Frontier, 1869–1880* (Albuquerque: University of New Mexico Press, 1996); and Sara Deutsch, *No Separate Refuge: Culture, Class, and Gender on the Anglo-Hispanic Frontier in the American Soutwest, 1880–1940* (New York: Oxford Press, 1987).

91. MacDonald, "Hispanic, Latino, Chicano, or 'Other,'" 396.

92. Yohn, *A Contest of Faiths,* 171–172.

10

Struggling for Voice in a Black and White World: The Lumbee Indians' Segregated Educational Experience in North Carolina

Heather Kimberly Dial

Lumbee Indians are a tribe of nonreservation[1] state-recognized Native Americans in Southeastern North Carolina. They are a unique indigenous people as evidenced by their unusual history and cultural origins. The Lumbee have historically lived in southeastern North Carolina, and their territory covered a large region, now known as Robeson, Hoke, Cumberland, and Scotland counties. The Lumbee tribe has grown from a few hundred who lived in this area at European contact, to 57,868, according to the 2000 census.[2]

This synthesis of secondary literature analyzes the historical development of segregated schools for Lumbee children in North Carolina. Native American researchers have documented the oral histories of the boarding schools, but the Lumbee experience has been overlooked.[3] This chapter adds to the body of historical literature on segregated educational experiences[4] and also to research on Indian Education.[5] It also fills a gap in the academic literature on state-supported segregation, which is currently limited to African American and Hispanic American populations.

Origins and Culture

Due to their efforts to define their cultural origins to non-Lumbee groups, the Lumbee people have been known by many names, such as Croatan Indians of Robeson County, Indians of Robeson County, and Cherokee Indians of

Robeson County.[6] The first attempt to trace the origin of the Lumbee drew from the history of the Lost Colony[7] and resulted in their designation as Croatan Indians of Robeson County. The Lost Colony theory holds that Lumbee Indians are a product of intermarriage between the inhabitants of the European colonist John White's failed colony and the Hatteras Indians of Roanoke Island,[8] and this theory is supported by oral history among the Lumbee.[9]

In 1913, the Lumbee were designated as the Cherokee Indians of Robeson County[10] on the basis of Robeson Senator Angus W. McLean's theory that they were descendants of the Cherokees who resisted Andrew Jackson's infamous removal[11] of Indians from the east.[12] Lumbee oral tradition also supports this theory.[13]

Other theories claim that the Lumbee originated from various Siouan tribes of North and South Carolina, such as the Waccamaw and Cheraw, or the Algonquian or Iroquoian Indians.[14] The Lumbee also explain their origin with reference to the possibility that they might have descended from Siouan tribes. There is a theory that the Lumbee are the remnants of the Cheraw Indian tribe; one group of Cheraw Indians came to Drowning Creek (Lumbee River) in present-day Robeson County and found solace and security in the swamps during the troublesome time of initial contact with Europeans.[15] Karen Blu[16] reasons that there may have been remnant groups of Indian tribes who found safety in the area of Robeson County and then intermarried, forming a single tribe that may have included non-Indians.[17]

Various theories demonstrate the richness of the Lumbee Indians' origins. In North Carolina, the Lumbee Indians historically experienced a very strange world where the major social, political, and economic forces were framed monochromatically by segregation in black and white. In nineteenth-century North Carolina, with the case of the Lumbee Indians, if one is neither white nor recognized as an Indian, then one is relegated to the status of black. This was inconsistent with the Lumbees' identification as Native American. An example of this denial of Lumbee Indian identity is evidenced by the plethora of labels North Carolina state government officials have given for the Lumbee, which include "'mixt crew,' 'free Negroes,' 'free persons of color,' 'mulattoes,' or 'mixed-blooded people'"[18]

The geographical area of North Carolina that the Lumbee ancestors called home was characterized by land largely considered undesirable by European settlers for a number of reasons. It mainly comprised swampland, woodland, and sandy soil, and the area was difficult to traverse and was unsuitable for farming.[19] The Lumbee homeland was part of the Cape Fear Valley area. Lying between North and South Carolina, it was bordered on the northeast by the Tuscarora Indians, who controlled the land between the Roanoke and Neuse rivers.[20] The presence of the powerful Tuscarora helped reinforce the seclusion of the Lumbee Indians in the Cape Fear Valley region, which enabled their unique Indian culture to flourish without much disruption. When English colonists defeated the Tuscarora in 1713, and the remaining Tuscarora finally abandoned their land in 1803, the area became open to English settlers.[21] The

Scottish Highlanders were the first immigrants to settle in the Cape Fear Valley region, and their arrival was a challenge to the peaceful isolation of the Lumbee Indians.[22]

Over time, the Lumbee Indians adopted the English language and a number of English customs. They grew beards, lived in European style homes, ran farms, owned slaves, and attended religious services.[23] Lumbee Indians adopted a form of Christianity associated with the Methodist and Baptist sects.[24] Due to their adoption of European customs and lifestyle, Lumbee Indians lost all vestiges of their original language, but maintained a unique dialect of English. Today, there are variations in the dialects of Lumbee English, marked by generational and regional differences.[25]

During the Indian Removal, from 1814 to 1858, the Lumbee, unlike other Indians, did not live in a tribal manner, but owned land individually. As a result, they were not considered Indians, so they posed little threat to the dominant culture.[26] The Lumbee existed in the isolation of the swamp and were considered insignificant until the political climate in North Carolina began to change in 1835 with the revisions of the state constitution.

Lumbee Resistance: Reclaiming Identity through Internalized Racism

Prior to 1835, the Lumbee community coexisted with the white community; they voted and attended school and church with whites.[27] Although a few Lumbee children who had fair skin and looked white[28] attended white schools that had been established in the Indian communities, most attended what were called "subscription schools" built by the Lumbee themselves.[29] (Subscription schools are defined as schools where the students paid the educators to attend school.) However, in 1835, the North Carolina Constitutional Convention decided, by a vote of 64 to 55, to disenfranchise free nonwhites.[30] The enactment of what were known as the Free Negro Codes entailed that: "[N]o free Negro, free mulattoe, or free person of mixed blood, descended from Negro ancestors to the fourth generation inclusive (though one ancestor of each generation may have been a white person) shall vote for members of the Senate or House of Commons."[31]

As the 1835 North Carolina Constitutional Convention did not refer specifically to the Lumbee, and as there was no official acknowledgement of them or their identity as Indians at this time, state legislators and policy makers assumed that this nonwhite population could be included in the category of "free Negro, free mulattoe, or free person of mixed blood."[32] This revision of the constitution, in effect, disenfranchised the Lumbee Indians, denied their Indian heritage, and breached their undisturbed isolation. One of the ways the Lumbee reacted to this denial of their heritage was to resist this classification with African Americans.

Cultural hegemony and colonization creates a social hierarchy wherein the dominant culture depends on the domination and oppression of non-dominant populations to reaffirm its status.[33] Such was the case in the South. In North Carolina, and all over the South, slavery, the Civil War, and Reconstruction were characterized by a black and white caste system of segregation and racial hierarchy. Enacted and enforced by a socio-economic and political structure wherein whites were the "upper caste," segregation and racial hierarchy became deeply entrenched social norms.

In this system, whites considered themselves superior because of their racial, cultural, political, and socioeconomic dominance.[34] Within the common hierarchy, blacks represented the "lower caste," and a number of laws regulated intimacy and contact between castes. The laws restricted the freedom of blacks and reinforced the inequality between the castes.[35] Regulation of the physical and social distance between castes served to preserve the "purity" of the "upper caste" from the "impurity" of the "lower caste". Prior to the Civil War, blacks were controlled by socioeconomic and political structures.[36] After the war, the so-called Jim Crow laws were enacted to control them.[37] These leaks limited their new freedom, and segregation replaced physical control. Blacks had minimal political power, could not vote, were excluded from public education, and had limited economic opportunities.[38]

Within this system, the Lumbee represented a unique manifestation of a caste within an otherwise black and white hierarchy and struggled to define themselves. Although they acknowledged they were not white, they knew also that they were not black. How were they to fit in this caste system? The issue was settled for the Lumbee by the North Carolina Constitutional Convention, which reclassified their ethnicity by including them in the category of "free Negro, free mulattoe, or free person of mixed blood."[39]

The Lumbee Native American identity has never been questioned internally by the tribe. According to Tatum,[40] identity is a complex concept "shaped by individual characteristics, family dynamics, historical factors, and social and political contexts."[41] In this sense, one's identity is a reflection of the world that surrounds one. The historical, social, and political factors that shaped the world of the Lumbee also helped foster the development of their Indian identity and an animosity toward blacks. Willis[42] provides a depiction of the climate of division among Indians and blacks in the eighteenth century, which whites helped foster in the southeast in the colonial era. Some Indian tribes of the southeast, specifically the Catawba Indians, were already hostile to blacks because of the competition for trade. But Willis argues that whites in the colonial era found themselves in a problematic position because they had two exploited groups—black slaves and Indian tribes—that could potentially challenge their racial supremacy.[43]

Whites feared the alliance of blacks and Indians, and endeavored to insulate these groups from each other. Interaction between these populations was first prevented by forbidding intermarriage and trading; next, Indian slavery was limited to forestall an alliance of Indian slaves and black slaves.[44] By prohibiting Indian–black interaction, whites believed they could prevent black slaves' flight into Indian country for freedom. To further prevent a black–Indian alliance and to maintain slavery, whites employed Indians to ferret out runaway slaves and Maroons.[45] Whites also attempted, as James Glen the governor of South Carolina explained, to "creat [sic] an aversion in them [Indians] to Negroes."[46]

They caused conflict between Indians and blacks by promoting fear, suspicion, and hatred between the two groups. Although there is no record of the tactics the whites used to create a rift between blacks and Indians, it is known that they spread rumors that blacks were responsible for the small-pox epidemic that devastated Indian tribes, and that they endeavored to ruin trade for the tribes. Whites also rewarded black slaves with freedom for fighting Indian tribes. This helped to add to the animosity that blacks had against Indians for their efforts as slave catchers and in fighting slave insurrections.[47] Thus, whites worked to develop enmity between blacks and Indians, because blacks viewed Indians as threats to their freedom and Indians viewed blacks as enemies.

While blacks were being controlled through segregation, Native Americans were subject to the Indian Removal Act, which resulted in their relocation from their homelands to the west. The government's failure to recognize the Lumbee as an Indian tribe was an advantage in that it protected them from Indian Removal. White supremacy left little room for negotiation; an acknowledged Lumbee presence would certainly challenge the black and white hierarchy. Prior to 1835, the Lumbee challenged this system by laying claim to an equal status with whites in voting, and attending school and church with whites.[48]

Lumbee based their assertion of equality with whites on the fact that when they encountered each other, they had already adopted the European lifestyle.[49] Culturally, Lumbee had long identified themselves with whites. During the colonial period, the Lumbee were not considered Indian, as the definition at that time was cultural rather than ethnic-racial as it was during the Civil War and Reconstruction. Due to the earlier cultural definition of Indian, the Lumbee no longer lived in a manner that aligned with the white perception of Indian tribal manner.[50]

As was the case with other Native American cultures, Lumbee culture was not static, but was constantly changing, depending on contact with Europeans. For the Lumbee, adoption of the English language and English customs represented a necessary accommodation to adapt and survive as a people alongside the European settlers. This contact *shaped* their Indian culture; it did not *eliminate* it. Historically, however, the dominant society has been intolerant of the realities of Native Americans such as the Lumbee, whose history challenges the continuity of what is thought to be Native American culture—a culture that is traditional and unchanging.[51]

During the Civil War and Reconstruction period in North Carolina, it was not the Lumbee Indians' cultural differences or similarities with whites, but their phenotype that changed the status quo in Lumbee–white relations. Southern whites during this time based equality and acceptance on ethnicity and race.[52] The Lumbee appeared racially mixed to the whites whom they encountered. This appearance gave rise to the labels given to the Lumbee, including, "mixt crew," "free Negroes," "free persons of color," "mulattos," or "mixed-blooded people."[53] The Lumbee suffered from prejudice and racism because they were nonwhite, and they defied the stereotype of what was considered Indian, which had been fixed in the minds of the whites. The Lumbee skin tones ranged from fair to dark.[54] Similarly, eye colors included brown, black, blue, green, and hazel.[55] Many Lumbee had European names due to acculturation and the various family heritages within their community.[56]

The most important challenge to their identity as Indians was Lumbee inter-marriage with members of other Indian communities as well as white and black communities. Although Native Americans are not all alike, the differing pheno-types within the Lumbee community were a problem for whites in their classi-fication of white and nonwhite because of their conceptions of blood—purity and impurity. The "one drop rule" was the definition of blackness in the South; it meant that one drop of black blood made a person racially black. The rule enabled whites to deny the Lumbee Indian identity because of the Lumbee intermarriage with blacks.

"Indianness"

There are three central components of "Indianness," or what it means to be Indian. Aspects of identity shared by all tribal Indians are blood and descent, relation to the land, and a sense of community.[57] Other external cultural iden-tity markers, including language, participation in cultural activities, dress, phys-ical features, consumption of Indian foods, and a particular lifestyle, are flexible and can be lost and regained. External elements of culture can be invented or reinvented, but the core of "Indianness" is descent, relationship to the land, and a sense of community.

Most significant in this idea of Indianness is bloodline descent. Indianness, in terms of blood and descent, has been defined differently by Indians and by the dominant culture. Herring[58] explains that "identity issues for North American Indians—including questions of mixed blood and full blood—stem from attitudes and ideas fostered by the majority European American culture and government. Before the "White man's coming," marriage across tribal and clan lines was common, and the offspring were not marked as mixed blood. Neither was tribal membership based on blood quantum or degree of accul-turation."[59]

Native Americans did not define Indianness in terms of blood quantum[60] but kinship.[61] Kinship was not limited to individuals who shared a common biological descent, but was extended to include other clans and nations.[62] It was not limited to biological reproduction of offspring, but allowed for indi-viduals and groups to become part of the Native American people "through naturalization, adoption, marriage, and alliance."[63] Indianness/identity involved inner qualities, beliefs, and the social action that unified them as a people, cohesiveness beyond mere blood. In contrast, the constructs of the biracial caste system developed by whites center on identity in matters of race, not culture.[64]

Before Indian Removal, Indianness had little to do with race; rather, kinship and maintaining kinship were important to Indian identity. Today, kinship is still important to Lumbee identity. The Lumbee tribal membership criteria require that an individual be able to trace descent from "persons who were identified as Indian on the source documents . . . and their direct descendants" (these documents include census records and the Croatan School Attendance List, which was the Lumbee school roll).[65]

Although Lumbee tribal membership specifically refers to blood descent, the community historically incorporated another way of establishing membership: through kinship. Fogelson[66] describes kinship among Indians before European contact as a "peoplehood" a network of communities wherein they had "reciprocal rights and duties toward one another and shared a collective sense of community."[67] The Lumbee sense of kinship is evidenced in members' efforts to "adopt" others by making them "Indian too" through marriage. However this sense of adoption and kinship is not always the case with Lumbee intermarriage with blacks.

Intermarriage with blacks was taboo in the Lumbee community. According to Pierce, "While most things could be forgiven, marrying black could not and young people were kept away from any contacts with Blacks. Individuals who married Blacks were forced to leave the community. There was no similar prohibition against marrying Whites within the Indian community, but it did not occur often, particularly toward the end of the century, because Whites would drive out the couple."[68]

Native Americans' definitions of Indianness differed from those of Europeans. Before they came into contact with Europeans, it was of little importance for American Indians to define who were Indians, because all inhabitants were Indians. When they came in contact with Europeans, they defined themselves with regard to other Indian tribes, most of whom referred to themselves as the "real people" and to other tribes as nonhuman. Europeans defined American Indians as a group, irrespective of tribe, and it took time for American Indians to view themselves in this manner as a category different from Europeans.[69] As Fogelson shows, European Americans defined Indian identity based on external characteristics or racial traits, such as physical features—skin color, hair, eye color—and the biophysical nature of the blood.[70]

The European focus on the biophysical nature of blood in defining racial identity taken into account the purity of blood from the point of view of genetic and racial homogeneity. This concern with blood purity can be traced to a "myth of blood,"[71] which suggests that blood permanently carries evidence of heredity of a people.

This understanding of blood purity gives another context for understanding white supremacy and the perception and treatment of nonwhites as inferior. It also helps in understanding the classification of Negros and Indians as individuals with "one-sixteenth part of 'Indian blood' or 'black blood'—that is, when one of their sixteen direct ancestors (great-great-grand-parents) was a Negro or an Indian."[72] This classification was the Spanish and Portuguese response to the desire for a logical order in referring to the European, Native American, African, and racial mixtures of these populations that had emerged in the new caste systems in the American colonies.[73] Specifically the definitions of mestizo, mulatto, and zambo were given to classify the new racial mixtures.[74] Descendants of Europeans and Indians were referred to as mestizo, those of Europeans and Africans as mulatto, and those of Africans and Indians as zambo.

In North Carolina, the social construction of race positioned the black cultural population as the "pariah race";[75] being associated with this cultural population lowered one's status. In 1715 North Carolina passed an antimiscegenation law prohibiting the intermarriage of whites and blacks.[76] Later, Lumbee disenfranchisement, the Jim Crow laws, and laws against intermarriage of whites with blacks served to control the black population and protect white supremacy. The segregation hierarchy, along with definitions of culture, race, ethnicity, and Indianness, provides a context for understanding the impact of Lumbee Indians' classification as black. Whites did not acknowledge a third category that might have served as a challenge to the biracial hierarchy. For the Lumbee, the effect of the biracial hierarchy was both to deny their Indian identity and to relegate them to the lower caste.[77]

The Lumbee resisted this categorization and hierarchy because it relegated them to second-class citizenship and denied their previous identity as equals of whites, an identity that had been developed through attendance at common churches and schools, and through voting with whites.[78] The Lumbee reified their Indian identity through internalized racism, by an assertion of Indian heritage, and by embracing segregation.[79] (A nondominant population's attempt to come to terms with racism by oppressing a similar group—excluding, delegitimizing, and demeaning it, distancing, and "othering" it—is known as internalized racism.)

Through their efforts to assert their Indian identity and impose self-segregation and internalized racism, the Lumbee endeavored to distinguish themselves from blacks. Some scholars[80] have hinted that these efforts were aimed at avoiding their being classified as black: that is, if the Lumbee could not become white, they would settle for being Indian.[81] However, research shows that a few members of the Lumbee community were identified by U.S. census takers as white, and there is evidence that some who left the community passed as white because of their lighter skin.[82]

The Lumbee distinguished themselves as "not black"[83] by asserting their Indianness, despising the cultural population of blacks, and accommodating white values.[84] It may be that by buying into this system of cultural hegemony and colonization, the Lumbee attempted to ally themselves with the dominant class by engaging in the oppression of blacks[85] and preventing themselves from being considered black.[86] The Lumbee attempt to distance themselves from blacks ultimately helped whites to create a rift between Indians and blacks. It was advantageous for whites to cause disharmony among these lower castes because of the danger of these populations uniting and working together to challenge white supremacy. Nat Turner's 1831 violent revolt against slavery and whites represented a radical and bloody resistance that substantiated their fears. The Lowrie War provides another example of violent resistance to the injustice of the caste system.[87] From 1864 to 1872, the Lowrie War involved a band of Indians, blacks, and poor whites led by an Indian, Henry Berry Lowrie, who engaged in violence to combat white tyranny in the South. Members of the Lowrie Band were modern-day Robin Hoods who used guerrilla tactics to avenge the injustice done to them and to other nonwhites in Robeson County.

After the Civil War, there were friendly relations between the Lumbee and their black neighbors. There was social and political unity as the example of the Lowrie Band of freedom fighters shows. However, the Lumbees' internalized racism strained the relations, because blacks then viewed the Lumbee as allying with whites rather than resisting segregation.[88] The Lumbee sought a position on the periphery of "white supremacy."

The internalized racism among the Lumbee benefited whites who sought power over both Lumbee Indians and blacks. It developed in the face of white oppression,[89] causing the Lumbee to simultaneously oppress blacks.[90] Instead of uniting in the struggle against oppression, the Lumbee worked against blacks by participating in the status quo of white domination.[91]

Segregated Lumbee Schools

After the Civil War, North Carolina again revised the state constitution in 1868 and 1875 so that it would reflect the policy to impart education to blacks, Indians, and whites. However, the revised constitution spoke only of white children and "colored" children being taught in separate schools, and did not acknowledge Native cultures in any way: "And the children of the white race and the children of the colored race shall be taught in separate public schools, but there shall be no discrimination made in favor of, or to the prejudice of, either race."[92] In 1869 the Freedman's Bureau[93] established schools for blacks in North Carolina, and in 1877 a normal school for blacks was established in Fayetteville.[94] The Lumbee refused to attend these black schools.

During Reconstruction the Lumbee, aided by the efforts of the historian and legislator Hamilton McMillan, sought not only their own schools, but also recognition of their Indian status.[95] McMillan, a historian from Red Springs, was the Robeson County representative and was well acquainted with the Lumbee Indians.[96] McMillan, took a personal interest in conducting, over several years, historical research on the origin of constituent Lumbee Indians. He sponsored legislation in 1885 to give Lumbees state recognition, entitling them to their own separate schools.[97]

The legislation McMillan brought before the North Carolina State General Assembly had two provisions: (1) to provide for the legal recognition of the Lumbee as an Indian people giving them the name Croatan Indians; and (2) to provide separate schools for them. In 1885 North Carolina passed the first state law regarding Indians that addressed education. It stated:

Whereas, the Indians now living in Robeson county claim to be descendants of a friendly tribe who once resided in eastern North Carolina on the Roanoke river, known as the Croatan Indians; therefore, The General Assembly of North Carolina do enact:
Section 1. That the said Indians and their descendants shall hereafter be designated and known as the Croatan Indians.
Section 2. That said Indians and their descendants shall have separate schools for their children, schools committees of their own race and color, and shall be allowed to select teachers of their own choice, subject to the same rules and regulations as are applicable to all teachers in the general school law.[98]

Although this act refers to the Indians in Robeson County, it included the Lumbee Indians from Hoke County because Hoke County was carved out of Robeson County in 1911. Croatan Indians who lived in counties outside Robeson as Scotland, Richmond, and Cumberland County also gained schools of their own.

The Croatan Indian Recognition Act of 1885 provides evidence of the Lumbee assertion of their Indian identity through their embrace of racial segregation. Beyond a refusal to attend schools with blacks, the racism of the Lumbee is evident in their attitude toward blacks. Hamilton McMillan, in communication with O. M. McPherson explained: "Since their recognition as a separate race they have made wonderful progress. Their hatred of the Negro is stronger than that entertained by Caucasians."[99] This hatred was another effort to disassociate themselves from blacks. When the supervisor of Indian Schools, Charles F. Pierce, was sent from the U.S. Department of the Interior to gather information about the Lumbee, he described them as follows: "They do not associate with the Negro race, looking upon them in about the same way as to [sic] do the Whites of their community."[100] The Lumbee's view of blacks and the effort to not associate with them mirrored the white oppression of blacks. Racism is also evident in the efforts of the Lumbee to advocate legislation in 1889 and 1911 to prevent African American children from attending their schools.[101]

In 1921 the North Carolina General Assembly passed legislation that established school committees to control admission to Indian schools, recommended teachers and principals for hiring and firing, and brought in Indian votes for key elections.[102] One committee established by the Lumbee through this legislation was a "blood committee," in existence until desegregation in 1954. It screened students who sought to enroll in the public schools and the normal school.[103] The committee was created to maintain the racial enrollment of the Indian schools, and it screened out students whose ancestors had black blood. This act gave the Lumbee the power to decide who attended their schools: no African Americans and no individuals of black mixed heritage, such as the group known as the Smilings, were admitted.

Dial and Eliades[104] explain that the Smilings were "the product of miscegenation [who] . . . migrated to Robeson County from the area of Sumter, South Carolina, after World War I."[105] They were referred to as such because Smiling was not a Lumbee family name[106] and the group could be easily differentiated. The predominant family names for the Smilings were Smiling and Epps.[107] A school was created for this separate group of Indians when they were denied admission to the Lumbee Indian schools. There were then four separate schools in Robeson County based on the designations—one each for blacks, whites, Indians, and the Smilings.[108] From the separate schools, we might conclude that the Lumbee were prejudiced against Indian groups that could not clarify their Indian heritage, such as the Smilings, and that these groups were not accepted fully into the Indian community. One could also argue that the Lumbee appropriated the one-drop blood distinction from whites.

Another example of Lumbee resistance and internalized racism can be seen in their opposition to the desegregation of their Indian schools. The Lumbee viewed the schools as a way to preserve their distinct cultural traditions which they feared would be lost through desegregation.[109]

Segregated Native American Schools and the Lumbee

Researchers have studied the education of Native American tribes with federal support.[110] The federal government supported these tribes' education between the late 1800s and the late 1940s, after which it contracted with the states for the education of these tribes. Federally supported tribal education was imparted by boarding schools administered by the Bureau of Indian Affairs to hasten Indian cultural transformation through education and to assimilate them into the mainstream.[111] The belief was that once Indians were assimilated they would be less dependent on the federal government. Boarding schools were located away from reservation lands to isolate students from their parents, their language, and their culture to hasten assimilation. Some students were seized and taken to the schools. In other cases parents were coerced into sending their children to these schools. The schools emphasized forced assimilation, harsh discipline, hard labor, and vocational training in their efforts to "kill the Indian to save the man."[112] Students were exposed to physical, emotional, and even sexual abuse.[113] The boarding school movement began in 1879 with the opening of the first off-reservation boarding school—the Carlisle Indian Industrial School, in Carlisle, Pennsylvania—and the assimilationist agenda of the boarding schools continued until the 1930s.[114] The educational experience of federally recognized tribes is different from that of the Lumbee. While boarding school experience is well documented, Lumbee Indian educational experience is missing from American Indian education literature.

Native American students were also educated in public schools in view of the federal termination[115] of support to tribes in the late 1940s.[116] The termination of support and allotment[117] of Indian lands were additional assimilationist efforts to change the Indians. Education of Indian students from these tribes with termination status then became the responsibility of the states.[118] Federal funds for Indian education were allocated to state schools to compensate them for the education of Indian students. For example, when Oklahoma became a state at the turn of the twentieth century, the state constitution defined blacks as anyone with black blood, and whites as everyone else, including Native Americans.[119] Thus, Indian students did not attend segregated schools, but attended schools with whites. Although the states were responsible for these students' education, the federal government provided the funding for it.[120]

Current historical literature does not document the state-supported education of tribes such as the Lumbee. A major study of American Indian education by Fuchs and Havinghurst has documented that the Lumbee are a unique case because they are a nonreservation tribe and historically had a separate school system within North Carolina's educational system.[121]

In North Carolina, segregated Indian schools varied by tribe. Tribes included the Eastern Band of Cherokee, Coharie, Person County, Waccamaw Siouan, Haliwa-Saponi, and the Lumbee. The education of the Eastern Band of Cherokee Indians in North Carolina was the responsibility of the federal government since they are a federally recognized tribe, but the government initially inconsistently provided for their education.[122] Schools for the Eastern Band of the Cherokee Indians took many forms during the eighteenth and nineteenth centuries. For example in the eighteenth century, Quaker missionaries established day schools and a boarding school in Cherokee.[123] The education these schools provided was deculturalizing to the extent that the goal was to kill Cherokee language and culture.

In 1885 the Lumbee received state recognition as Indians and, as a result, they secured their own schools in Robeson and Hoke counties. This reaffirmed the racial identity of the Lumbee. The Lumbee went willingly to these schools. They had lived separately from blacks and whites and had sought to establish their own segregated schools for 50 years. The Lumbee schools came to represent more than an affirmation of Indian identity to the Lumbee people. As Sider explains, "More than being a core feature of Lumbee political organization, the separate elementary and high schools have been crucial to an Indian cultural and social life: schools and churches are the places where Indians become connected to other Indians. They are not simply the center of the community; in fundamental ways they are the community."[124] For the Lumbee, the segregated Indian schools served as places where they bound themselves together both as a tribe for political action and as a community. Furthermore, the schools enabled the Lumbee to transmit their cultural values, traditions, and heritage. The schools and the cultural identity of the Lumbee were intertwined.

Education became a defining cultural characteristic of the Lumbee. Their struggle for schools of their own helped reify their Indian identity in opposition to their previous categorization as blacks. Indian schools became a place of cultural, social, economic, and political agency for them. The involvement of Lumbee teachers, administrators, parents, and community members in running the schools perpetuated Lumbee ideals, beliefs, history, heritage, and culture. The schools were central to the social life of the Lumbee, and as my research participants explained, there were only two places that members of the community could go to socialize: the church and school. Education and teaching in the Indian schools increased Lumbee economic opportunities. Teaching was secure white-collar work and provided the Lumbee an alternative to farming. It represented a new career opportunity wherein the Lumbee had access to education and could become educators themselves in the schools. Thus, education and teaching became synonymous for the Lumbee, just as their excellent skills as farmers had once defined them. The Lumbee were perhaps the most educated of all Native American tribes because of the impact of the University of North Carolina-Pembroke, the former Indian Normal School, and the Lumbee segregated Indian schools.[125]

The schools also provided political agency for the Lumbee because it was through the establishment of separate schools that they were acknowledged as Indians and were able to exert some power.[126] Their agency was demonstrated by their control over who was admitted to their schools and who taught in them.[127] Furthermore, the Indian schools developed Lumbee leaders who would advocate for the Lumbee socially, politically, and economically.[128] Finally, the Indian schools were also tied to Lumbee identity because one of the criteria of tribal membership was proof of descent from individuals who attended the segregated Indian schools.[129]

Although the 1885 act provided separate schools for the Lumbee Indians in Robeson County, the Lumbee remained at a disadvantage. The act provided funds to pay teachers and instructed the treasurer to distribute the school funds evenly, and therefore the Lumbee Indians received their proportional share of the school funds. However, the law did not provide funding for building separate schools or for the education of Indian teachers. Representative McMillan[130] brought forth this concern to the legislature on behalf of the Lumbee, resulting in the establishment of an Indian Normal School[131] for teacher training. The 1885 act provided $500 to pay school instructors, but did not provide for building construction; hence, the Lumbee had to build the school on their own. In 1887 Lumbee school trustees donated $1,000 worth of materials and labor to build the Croatan Normal School.[132] Although the school was initially inadequately funded, the legislature granted an additional $1,000 in 1889 and the school continued to operate.[133]

In 1890, two years after an unsuccessful request for federal funding, W. L. Moore, an activist for the Lumbee, contacted the Office of Indian Affairs in Washington, D.C., to ask for funding for the education of the Lumbee. Commissioner of Indian Affairs T. J. Morgan responded to Moore based on the information Hamilton McMillan had provided to him about the Lumbee:

> It appears from his statement that this band is recognized by the State of North Carolina, has been admitted to citizenship, and the state has undertaken the work of their education.
>
> While I regret exceedingly that the provisions made by the State of North Carolina seem to be entirely inadequate, I find it quite impractical to render any assistance at this time. The Government is responsible for the education of something like 36,000 Indian children and has provisions for less than half this number. So long as the immediate wards of the Government are so insufficiently provided for, I do not see how I can consistently render any assistance to the Croatans or any other civilized tribes.[134]

Morgan's statement reveals the prejudice of the federal government in dealing only with federally recognized tribes. But while Morgan proclaimed that the education of Indians was insufficiently provided for, he refused to provide funding.[135] The Lumbee (Croatan) tribe was denied financial assistance because they were considered a civilized tribe, and the purpose of federal education for American Indian tribes was to "civilize" them.

Charles F. Pierce in 1912 echoed Morgan's words in denying federal assistance to the Lumbee on the basis of U.S. Indian School policy decisions:

> At the present time it is the avowed policy of the government to require the states having an Indian population to assume the burden and responsibility for their education, so far as possible. North Carolina, like the state of New York, has a well organized plan for the education of the Indians within her borders, and there does not appear to be any justification for any interference or aid on the part of the government in either case, especially in a prosperous community like Robeson County, North Carolina.[136]

In 1900, 1912, and 1933, the Lumbee sought federal funding for an Indian school that would offer more advanced courses than those available at the Indian Normal School, since the other state schools were not open to them, but to no avail.[137] In 1933, the federal government insisted that it had no treaties with the Lumbee to provide for their education and were not obligated to provide school facilities.[138] Lumbee people had limited means of making a living—as farmers, clergy, or educators in the segregated Indian schools. Separate schools gave the Lumbee access to secure jobs that paid regularly and did not hinder them from continuing to farm or to preach. This enabled the development of a Lumbee middle class of educators.[139]

Lumbee School System in North Carolina

The Lumbee Indians had separate primary and secondary schools in both Robeson and Hoke counties, where they were part of a tripartite (white, black, and Lumbee) school system in each county. This system was a result of the 1875 revision of the North Carolina Constitution to provide separate schools for white and nonwhite children. The constitution was further amended in 1923 to include reference to the Lumbee, or Croatan Indians, as they were known at the time:

> The children of the white race and the children of the colored race shall be taught in separate public schools, but there shall be no discrimination in favor of or to the prejudice of either race. All white children shall be taught in the public schools provided for the white race, and all colored children shall be taught in the public schools provided for the colored race; but no child with negro blood, or what is generally known as Croatan Indian blood, in his veins, shall attend a school for the white race, and no such child shall be considered a white child. The descendants of the Croatan Indians, now living in Robeson, Sampson, and Richmond counties, shall have separate schools for their children.[140]

The Lumbee schools in Robeson and Hoke counties were both under boards of education that were white controlled. Furthermore, these Indian schools and the Indian Normal School were under the state supervision of the Division of Negro Education. This supervision and categorization under the Division of Negro Education was an affront to the Lumbee, and they lobbied to have the

Indian school records kept separate.[141] However, the legislation was approved to keep the entire records separate only at the county level and not at the state level.[142]

The Indian school system in Robeson County was the last of the three systems to be developed (white, black, and Indian).[143] Of these systems, Indian teachers were paid the lowest salaries and were the fewest in number.[144] The school buildings were smaller, and the quality of construction was not comparable with that of the white schools. Fewer textbooks were available to the students in the Indian schools than to white students. However, research shows that the conditions in black schools were worse than those in the white and Indian schools in Robeson County.

Ernest Hancock conducted a study of the three schools systems in Robeson County and found that the black schools suffered the most from inequities. There were 32 school buildings for Indians, including 5 of brick construction, compared with 3 brick structures among 55 school buildings for blacks. Hancock considered the buildings of the black schools old and poorly constructed compared with Indian and white schools. Most of the black schools were one-room schools with light and heat provisions. In evaluating the teaching resources and equipment, Hancock found that both Indian and black schools were poorly equipped compared with the better equipped white schools. All of the Indian and black schools and most of the white schools were overcrowded. The county attempted to address the overcrowding by manufacturing crude desks for the black and Indian schools, but Hancock did not find any such desks in the white schools.[145]

In his research, Hancock found that the school buses for the Indian and black students were old, in need of repair, and unreliable, resulting in students' inability to get to school on time and missing instruction. Fewer black students went to school by bus in comparison with their Indian and white counterparts. Among the faculty in the Robeson County schools, there were fewer Indian teachers: the breakdown was 181 white teachers, 94 Indian teachers, and 162 black teachers.[146]

The black schools in Robeson County had lower enrollments and monetary value than Indian schools, as shown by the Robeson County Board of Education Annual Statistical Report for the years 1951–1952. According to the report, black schools were valued at $738,040 with an enrollment of 3,319 students.[147] Indian schools were valued at $2,090,360 with an enrollment of 7,910.[148] The inequalities of resources between the Indian and black school systems with the same white leadership are comparable with the inequalities found in research on the segregated African American schools, which were the impetus for the *Brown v. Board* case.[149]

By 1970, all Indian schools in North Carolina had ceased to operate except for those in Robeson County that were among the last schools in the nation to be desegregated.[150] The Lumbee vehemently opposed desegregation in the county through protests and sit-ins at the former Indian schools. One reason the Lumbee fought segregation was internalized racism, which Massey[151] describes as Native American racism targeted at blacks.

Ironically, although the Lumbee were discriminating against blacks, their fight to retain segregated schools was their way of combating racism against themselves and an effort to protect their identity. The late Dexter Brooks, one of the first Lumbee appointed to serve as a superior court judge in Robeson County, explained in an interview that the schools protected the Lumbee:

> Indians avoided the worst aspects of racism during the Jim Crow[152] era by literally creating their own communities . . . many communities were self-contained . . . when I was a kid you could go weeks without coming in contact with a non-Indian . . . back then it was an advantage because you avoided all the racism.[153]

Schools enabled the Lumbee to preserve their distinct cultural traditions, which they feared would be lost through desegregation.[154]

The Lumbee model of schooling spread in North Carolina. In 1885, the state had two recognized tribes of Native Americans—the federally recognized Eastern Band of Cherokee and the state-recognized Lumbee Indians. The Lumbee Indians' successful struggle for state recognition as an Indian tribe set the precedent for other unrecognized Indian tribes in the state to validate their Indian identity and to secure their own schools.[155] Many unrecognized Indian tribes such as the Sappony, the Cohaire, and the Waccamaw-Sioux, endeavored to designate themselves as Croatan Indians in order to fund their Indian schools.[156] The Sappony Indians, historically known as the Person County Indians, successfully petitioned for state recognition as Indians in 1913 and were permitted a school of their own by the Person County Board—the High Plains Indian School.[157] Later the Coharie Indians followed suit by gaining recognition and securing their own schools in New Bethel and East Carolina.[158] The Waccamaw-Sioux Indians had a more difficult struggle for recognition and separate Indian schools, and did not gain Hickory Hill School until 1933.[159] Haliwa-Saponi Indians did not receive state recognition and have their own separate school until 1957.[160] The Haliwa Public School was closed in 1970 because of desegregation,[161] but it reopened in 2000 as the Haliwa-Saponi Tribal School.[162] The Cherokee, Cohaire, Sappony, Waccamaw-Siouan, and Haliwa-Saponi are not the only Indian tribes in North Carolina, but they are the ones that had separate schools prior to desegregation. Other Indian tribes in North Carolina include the Meherrin, who were recognized by the state as an Indian tribe in 1986, and the Occaneechi Band of Saponi Nation, who were the last Indian tribe to gain state recognition in 2002.

Lumbee Opportunities for Higher Education

The Indian Normal School, which is known today as the University of North Carolina at Pembroke, was once the only teacher-training institution in North Carolina for the Lumbee. The school enabled Lumbee to become teachers, and the first class graduated in 1928.[163] A full college curriculum was developed in 1930s, and the first college degrees were conferred in 1940, transforming the normal school into the Pembroke State College for Indians,[164] the first state-supported college for Indians in the United States.[165] Until 1945, only Lumbee Indians were allowed to attend the college, after which it was opened to all Native Americans.

Before 1942, the college offered only teaching degrees for the Lumbee. Thus if any Lumbee wished to pursue an educational degree other than teaching or attend graduate school, they had to leave the area and seek other colleges that would admit them. Although Indian boarding schools offered career opportunities other than teaching, they were not sufficient for members of the Lumbee community who wished to pursue higher degrees. North Carolina provided institutions of higher education for white and black populations, but these institutions by law were not open to the Lumbee Indians.[166] In order to learn trades, some Lumbee Indians attended the Carlisle Indian Industrial School. This is significant, because only members of federally recognized tribes were admitted, and the acceptance of the Lumbee in this school represents recognition that they were Indian.[167]

In North Carolina there were six public institutions of higher learning for nonwhites. Five of these universities were for African American students: Elizabeth City State University, Fayetteville State University, North Carolina Agricultural and Technical State University, North Carolina Central University, and Winston-Salem State University. The University of North Carolina at Pembroke was the only institution for Indian students. The other public institutions of higher education in North Carolina were for whites, and nonwhites by law were not allowed admission to these schools.

The Second Morrill Act of 1890 gave equitable funding to establish and maintain land grant institutions for teaching agricultural and mechanical arts for white and colored students. The act also funded North Carolina State University (NCSU), established in 1887 and opened in 1889 as the North Carolina College of Agriculture and Mechanic Arts.[168] NCSU, at this time, was the white college for teaching agricultural and mechanical arts.[169] The act permitted states to receive federal funding only if race or color was not a criterion for admission; otherwise, the states had to establish separate but equally equipped institutions for black students. The act is another example of segregation and higher education that impacted higher educational institutions in North Carolina.

However, explicit legislation was not always a factor in the segregation of higher institutions. In the case of the University of North Carolina at Chapel Hill, it was the university trustee policy to exclude black students. Although there was not a statute excluding black students, the policy had the same effect as law.[170] The policy changed in 1951, when the first black students were admitted.[171] In 1957 the language of racial exclusion was removed from the legislation for the public universities of North Carolina.[172] The new statement of mission read: "The primary purpose of Pembroke State College shall be the undergraduate education of the Lumbee Indians and other persons who may be admitted under uniform regulations of the Board of Trustees."[173]

Before 1951 the Lumbee, though they were taxpaying citizens of North Carolina, could not attend these public universities because they were neither white nor black. These institutions only recognized white or black, and most

Lumbee were reluctant to "pass" as black and preferred not to attend institutions for black students. Some Lumbee successfully gained admission to white colleges in North Carolina, such as Duke University, Wake Forest University, University of North Carolina at Chapel Hill, University of North Carolina at Greensboro, and Western Carolina University, in their pursuit of graduate degrees.[174] It is possible that the Lumbee who attended these white colleges "passed" for white and did not reveal their Indian identity for fear of rejection. Other Lumbee, and also many African Americans, however, were denied admission to white colleges in North Carolina and thus had to pursue their graduate education at universities outside of North Carolina, such as the University of Georgia, the University of Michigan, Ohio State University, University of South Carolina, and George Peabody College.[175] One former Lumbee student explained that at the University of Georgia he was documented on school admission records as white, because there was no category for Native American.[176] He never denied his Lumbee Indian heritage in gaining admission to the school, but the school classified him as white to admit him.

Conclusion

This history of Lumbee education adds to the American Indian education literature by documenting the educational experiences of a state-supported tribe that sought its own separate schools for complex social, cultural, and historical reasons. Within the confines of discriminatory state and federal legislation, the segregated Indian schools bound Lumbees together, both as a tribe for political action and as a community. Furthermore, their schools enabled the Lumbee to transmit their cultural values, traditions, and heritage. Their struggle for schools of their own helped reify their Indian identity. For them, Indian schools were a place of cultural, social, economic, and political agency.

Notes

1. The term *nonreservation* refers to the fact that the Lumbee have never lived on reservations, but rather owned their own land or shared common land. For a discussion of reservations, see F. M. Bordewich, *Killing the White Man's Indian* (New York: Anchor Books, 1996); A. J. Cobb, *Listening to Our Grandmothers' Stories: The Bloomfield Academy for Chickasaw Females, 1852–1949* (Lincoln: University of Nebraska, 2000); A. L. Dial and D. K. Eliades, *The Only Land I Know: A History of the Lumbee Indians* (San Francisco: Indian Historian Press, 1975); P. L. Marshall, *Cultural Diversity in Our Schools* (Belmont, CA: Wadsworth/Thomson Learning, 2002); J. H. Spring, *Deculturalization and the Struggle for Equality: A Brief History of the Education of Dominated Cultures in the United States* (New York: McGraw-Hill, 2001).

2. G. M. Sider, "Lumbee Indian Cultural Nationalism and Ethnogenesis," *Dialectical Anthropology* 1 (1976); S. U. Ogunwole, *American Indian and Alaska Native Population: 2000, Census 2000 Brief* (issued February 2002), http://www.census.gov/prod/2002pubs/c2kbr01-15.pdf (accessed March 4, 2004).

3. Cobb, *Listening to Our Grandmothers' Stories*; C. Ellis, *To Change Them Forever: Indian Education at the Rainy Mountain Boarding School, 1893–1920* (Norman: University of

Nebraska Press, 1996); K. T. Lomawaima, *They Called It Prairie Light: The Story of Chilocco Indian School* (Lincoln: University of Nebraska Press, 1994).

4. J. D. Anderson, *The Education of Blacks in the South, 1860–1935* (Chapel Hill: The University of North Carolina Press, 1988); H. S. Ashmore, *The Negro and the Schools* (Chapel Hill: University of North Carolina Press, 1954); H. M. Bond, *Negro Education in Alabama: A Study in Cotton and Steel* (Washington: Associated Publishers, 1939); D. Cecelski, *Along Freedom Road: Hyde County, North Carolina, and the Fate of Black Schools in the South* (Chapel Hill: University of North Carolina Press, 1994); D. E. Emerson, "Education to Subordinate—Education to Liberate: An Historical Study of the Dual Role of Education for African Americans, 1865—1968" (doctoral dissertation, North Carolina State University, 2003); P. A. Hessling, "To Be 'the Best School in Town': An Historical Study of Two Southern Elementary Schools" (doctoral dissertation, University of North Carolina at Chapel Hill, 1993); V. S. Walker, *Their Highest Potential: An African American School Community in the Segregated South* (Chapel Hill: University of North Carolina Press, 1996); A. V. Wilson, "Forgotten Voices: Remembered Experiences of Cross-over Teachers During Desegregation in Austin, Texas, 1964–1971" (doctoral dissertation, University of Texas at Austin, 1997); A. V. Wilson and W. E. Segall, *Oh, Do I Remember! Experiences of Teachers During the Desegregation of Austin's Schools, 1964–1971* (New York: SUNY Press, 2001).

5. Cobb, *Listening to Our Grandmothers' Stories;* Ellis, *To Change Them Forever: Indian Education at the Rainy Mountain Boarding School, 1893–1920;* E. Fuchs and R. J. Havinghurst, *To Live on This Earth: American Indian Education* (Garden City, NJ: Anchor Books, 1973); Lomawaima, *They Called It Prairie Light;* Lomawaima, "Educating Native Americans," in *Handbook of Research on Multicultural Education,* ed. J. A. Banks and C. A. McGee Banks (New York: Macmillan, 1995); Spring, *Deculturalization;* K. G. Swisher and J. W. Tippeconnic, III, eds., *Next Steps: Research and Practice to Advance Indian Education* (Charleston, WV: Appalachia Educational Laboratory, 1999); M. C. Szasz, *Education and the American Indian: The Road to Self-Determination since 1928* (Albuquerque, NM: University of New Mexico Press, 1999).

6. An act relating to the Lumbee Indians of North Carolina, North Carolina Session Laws, 1953, c. 874; An act to change the name of the Indians in Robeson County and to provide for said Indians separate apartments in the state hospital, Public Laws of North Carolina, 1911, c. 215; An act to provide for separate schools for Croatan Indians in Robeson County, Laws of North Carolina, 1885, c. 51; An act to restore to the Indians residing in Robeson and adjoining counties their rightful and ancient name, Public Laws of North Carolina, 1913, c.123.

7. *Lost Colony* refers to the colony established by Sir Walter Raleigh's charter to claim land in North America. The Lost Colonists were theorized to have joined the Hatteras Indians and intermarried with them. This is one of the theories of the probable origins of the Lumbee Indians. See also Dial and Eliades, *The Only Land I Know;* H. McMillan, *Sir Walter Raleigh's Lost Colony* (Raleigh, NC: Edwards and Broughton Company, 1888).

8. J. Lawson, A new voyage to Carolina; Containing the exact description and natural history of that country: Together with the present state thereof. And a journal of a thousand miles, travel'd thro' several nations of Indians. Giving a particular account of their customs, manners, & c. London: [s.n.], 1709, http://docsouth.unc.edu/nc/lawson/menu.html (accessed March 6, 2004); McMillan, *Sir Walter Raleigh's Lost Colony.*

9. L. Barton, *The Most Ironic Story in American History: An Authoritative Documented History of the Lumbee Indians of North Carolina* (Charlotte: Associated Printing Corporation, 1967); Dial and Eliades, *The Only Land I Know;* McMillan, *Sir Walter Raleigh's Lost Colony.*

10. *Cherokee Indians of Robeson County* refers to the belief of several scholars that in the early eighteenth century some Cherokee warriors who were part of the forces of Colonel John Barnwell decided to stay and join in with the Lumbee in present-day Robeson County on their return home from trying to quell a Tuscarora Indians uprising in 1711. See Dial and Eliades, *The Only Land I Know.*

11. *Indian Removal* refers to the time period of 1814–1858 when the government policy inspired by Andrew Jackson was to segregate Indians from whites through the forced military relocation of Indians from their homelands in the east to lands set aside west of the Mississippi River. See Spring, *Deculturalization.*

12. Angus Wilton McLean et al., *Lumber River Scots and Their Descendants; the Mcleans, the Torreys, the Purcells, the Mcintyres, the Gilchrists* (Richmond, VA: The William Byrd Press Inc., 1942).

13. A. L. Dial, *The Lumbee* (New York: Chelsea House Publishers, 1993); McLean et al., *Lumber River Scots and Their Descendants.*

14. Karen I. Blu, *The Lumbee Problem: The Making of an American Indian People* (Lincoln: University of Nebraska Press, 2001); James Mooney, *The Siouan Tribes of the East* (Washington, D.C.: GPO, 1894); John R. Swanton, "Probable Identity of the 'Croatan' Indians," in *U.S. Senate Reports, Siouan Indians of Lumber River* (Washington, D.C.: GPO, 1934).

15. James Hart Merrell, *The Indians' New World: Catawbas and Their Neighbors from European Contact through the Era of Removal* (Chapel Hill: Published for the Institute of Early American History and Culture, Williamsburg, VA, by the University of North Carolina Press, 1989).

16. Blu, *The Lumbee Problem.*

17. These tribes may have found Robeson County appealing because of the geographic isolation and fruitfulness of the land, the disputed border, and the presence of other native tribes in that area that they may have known through the trade route on the Lumbee River. The Native American tribes may have sought to settle in this area because they were seeking a haven from European diseases such as smallpox, were displaced by European settlers' encroachment upon their land, and were avoiding the Tuscarora War of 1711. This "peoplehood" of Natives and non-natives may have used English as a common language that replaced their native languages. See Blu, *The Lumbee Problem;* W. Wolfram et al., *Fine in the World: Lumbee Language in Time and Place* (Raleigh: North Carolina State University Humanities Extension/ Publications, 2002).

18. Dial, *The Lumbee.*

19. Dial and Eliades, *The Only Land I Know;* M. M. Maynor, "Violence and the Racial Boundary: Fact and Fiction in the Swamps of Robeson County, 1831–1871" (honors thesis, Harvard University, 1995).

20. Theda Perdue, *Native Carolinians: The Indians of North Carolina* (Raleigh: North Carolina Division of Archives and History, 1985).

21. Ibid.

22. Dial and Eliades, *The Only Land I Know.*

23. Barton, *The Most Ironic Story in American History: An Authoritative Documented History of the Lumbee Indians of North Carolina;* Dial and Eliades, *The Only Land I Know;* McMillan, *Sir Walter Raleigh's Lost Colony.*

24. Dial and Eliades, *The Only Land I Know.*
25. Wolfram et al., *Fine in the World.* .
26. Blu, *The Lumbee Problem*; Dial, *The Lumbee.*
27. Fuchs and Havinghurst, *To Live on This Earth*; Maynor, "Violence and the Racial Boundary; O. M. McPherson, "Indians of North Carolina: A Report on the Condition and Tribal Rights of the Indians of Robeson and Adjoining Counties of North Carolina" (Washington, D.C.: GPO, 1915).
28. Julian T. Pierce, *The Lumbee Petition* (Pembroke, NC: Lumbee River Legal Services, 1987).
29. McPherson, "Indians of North Carolina."
30. "Journal of the Convention, Called by the Freemen of North-Carolina, to Amend the Constitution of the State, Which Assembled in the City of Raleigh, on the 4th of June, 1835, and Continued in Session until the 11th Day of July Thereafter" (Raleigh, NC: Printed by J. Gales & Son, Printers to the Convention, 1835); Maynor, "Violence and the Racial Boundary.
31. "North Carolina Constitutional Convention," 98.
32. Ibid.
33. J. Comaroff and J. L. Comaroff, *Of Revelation and Revolution,* vol. 2, *The Dialectics of Modernity on a South African Frontier* (Chicago: University of Chicago Press, 1997); Marshall, *Cultural Diversity in Our Schools.*
34. J. H. Franklin, *The Free Negro in North Carolina, 1790–1860* (Chapel Hill: University of North Carolina Press, 1995); Wilson, "Forgotten Voices"; C. Vann Woodward, *The Strange Career of Jim Crow,* 3rd rev. ed. (New York: Oxford University Press, 1974).
35. Woodward, *The Strange Career of Jim Crow.*
36. Wilson, "Forgotten Voices"; Woodward, *The Strange Career of Jim Crow.*
37. Woodward, *The Strange Career of Jim Crow.*
38. Wilson, "Forgotten Voices"; Woodward, *The Strange Career of Jim Crow.*
39. "North Carolina Constitutional Convention," 98.
40. B. D. Tatum, *"Why Are All the Black Kids Sitting Together in the Cafeteria?" and Other Conversations About Race* (New York: Basic Books, 1997).
41. Ibid.
42. William S. Willis, "Divide and Rule: Red, White, and Black in the Southeast," *Journal of Negro History* 48, no. 3 (1963).
43. Ibid.
44. Ibid.
45. *Maroons* were fugitive slaves that set up their own societies and worked against slavery by attacking white settlements and through liberating slaves. Ibid.
46. Ibid., 165.
47. Ibid.
48. Fuchs and Havinghurst, *To Live on This Earth: American Indian Education;* Maynor, "Violence and the Racial Boundary."
49. McMillan, *Sir Walter Raleigh's Lost Colony;* McPherson, "Indians of North Carolina"
50. Dial and Eliades, *The Only Land I Know.*
51. James Clifford, *The Predicament of Culture: Twentieth-Century Ethnography, Literature, and Art* (Cambridge, MA: Harvard University Press, 1988).
52. Dial and Eliades, *The Only Land I Know.*
53. Dial, *The Lumbee.*
54. Blu, *The Lumbee Problem*; Bordewich, *Killing the White Man's Indian;* Maynor, "Violence and the Racial Boundary."

55. Blu, *The Lumbee Problem;* Bordewich, *Killing the White Man's Indian.*
56. McMillan, *Sir Walter Raleigh's Lost Colony;* McPherson, "Indians of North Carolina";
    G. M. Sider, *Lumbee Indian Histories:Race, Ethnicity, and Indian Identity in the Southern United States* (New York: Cambridge University Press, 1993).
57. R. D. Fogelson, "Perspectives on Native American Identity," in *Studying Native America: Problems and Prospects,* ed. Russell Thornton (Madison: University of Wisconsin Press, 1998).
58. R. D. Herring, "Native American Indian Identity: A People of Many Peoples," in *Race, Ethnicity, and Self: Identity in Multicultural Perspective,* ed. E. P. Salett and D. R. Koslow (Washington, D.C.: National MultiCultural Institute, 1994).
59. Ibid.
60. *Blood quantum* refers to the degree of Indian blood an individual possesses: one may have no Indian blood, or one may be a full-blooded Indian. This method of calculating the degree of Indian blood was evolved by the government as an administrative measure of Indian identity to determine which Indians were eligible for benefits and services. See Fogelson, "Perspectives on Native American Identity"; Gary A. Sokolow, *Native Americans and the Law: A Dictionary, Contemporary Legal Issues* (Santa Barbara, CA: ABC-CLIO, 2000).
61. Fogelson, "Perspectives on Native American Identity."
62. Ibid.
63. Ibid.
64. Dial and Eliades, *The Only Land I Know;* Fogelson, "Perspectives on Native American Identity."
65. Pierce, *The Lumbee Petition,* 233.
66. Fogelson, "Perspectives on Native American Identity."
67. Ibid.
68. Pierce, *The Lumbee Petition,* 149.
69. Russell Thornton, *American Indian Holocaust and Survival: A Population History since 1492,* 1st ed. (Norman: University of Oklahoma Press, 1987).
70. Fogelson, "Perspectives on Native American Identity."
71. The Bureau of Indian Affairs (BIA), an agency of the dominant culture and the federal government, defines Indianness through blood quantum. An Indian, according to the BIA, is a person whose blood quantum is at least one quarter Indian; see *"Bureau of Indian Affairs Answers to Frequently Asked Questions"* (BIA, DOI, 2004). However, another federal agency, the Census Bureau, relies on self-identification; see *"Facts on the American Indian/Alaska Native Population"* (U.S. Census Bureau, 2003); Ogunwole, *American Indian and Alaska Native Population: 2000, Census 2000 Brief;* Fogelson, "Perspectives on Native American Identity." Today, the responsibility for blood quantum falls on the individual tribal governments and can vary greatly from tribe to tribe.
72. Juan Comas, *Racial Myths* (Paris: UNESCO, 1958), 18.
73. Jack D. Forbes, *Africans and Native Americans: The Language of Race and the Evolution of Red-Black Peoples,* 2nd ed. (Urbana: University of Illinois Press, 1993); Winthrop D. Jordan, *White over Black: American Attitudes toward the Negro, 1550–1812* (Chapel Hill: Published for the Institute of Early American History and Culture at Williamsburg, VA, by the University of North Carolina Press, 1968).
74. Forbes, *Africans and Native Americans;* Thornton, *American Indian Holocaust .*
75. John U. Ogbu, *Minority Education and Caste: The American System in Cross-Cultural Perspective* (New York: Academic Press, 1978).
76. Franklin, *The Free Negro in North Carolina, 1790–1860.*

77. Blu, *The Lumbee Problem;* Dial and Eliades, *The Only Land I Know;* Maynor, "Violence and the Racial Boundary."

78. Fuchs and Havinghurst, *To Live on This Earth;* Maynor, "Violence and the Racial Boundary"; McPherson, "Indians of North Carolina."

79. bell hooks, "Revolutionary 'Renegades,'" in *Black Looks: Race and Representation* (Boston, MA: South End Press, 1992); M. M. Maynor, "People and Place: Croatan Indians in Jim Crow Georgia" (master's thesis, University of North Carolina at Chapel Hill, 2002).

80. Brewton Berry, *Almost White* (New York: Macmillan, 1963), G. B. Johnson, "Personality in a White-Indian-Negro Community," *American Sociological Review* 4, no. 4 (1939).

81. Berry, *Almost White;* Johnson, "Personality in a White-Indian-Negro Community."

82. Blu, *The Lumbee Problem.*

83. Ibid. See also Dial, *The Lumbee;* Maynor, "People and Place"; Maynor, "Violence and the Racial Boundary"; and Perdue, *Native Carolinians.* For more views on the Lumbee resistance to black racial classification.

84. Maynor, "People and Place."

85. An act to amend the laws of 1885 and 1887 so as to provide additional educational facilities for the Croatan Indians; citizens of Robeson County, North Carolina, Laws of North Carolina, 1889, c. 60; An act to empower the trustees of the Indian Normal School of Robeson County to transfer title to property of said school by deed to State Board of Education, and to provide for the appointment of trustees for said school, Public Laws of North Carolina, 1911, c. 168.

86. hooks, "Revolutionary 'Renegades'"; L. L. Lovett, "'African and Cherokee by Choice': Race and Resistance under Legalized Segregation," in *Confounding the Color Line,* ed. J. F. Brooks (Lincoln: University of Nebraska Press, 2002).

87. William McKee Evans, *To Die Game: The Story of the Lowry Band, Indian Guerrillas of Reconstruction (Iroquois and Their Neighbors),* 1st ed. (Syracuse, NY: Syracuse University Press, 1995); Maynor, "Violence and the Racial Boundary."

88. hooks, "Revolutionary 'Renegades'"; Lovett, "'African and Cherokee by Choice.'"

89. Dial and Eliades, *The Only Land I Know;* "North Carolina Constitutional Convention."

90. hooks, "Revolutionary 'Renegades'"; Lovett, "'African and Cherokee by Choice.'"

91. hooks, "Revolutionary 'Renegades'"; Lovett, "'African and Cherokee by Choice.'"

92. John L. Cheney, Jr., ed., *North Carolina Government, 1585–1979: A Narrative and Statistical History,* an updated edition of North Carolina Government, 1585–1974. (Raleigh: North Carolina Department of the Secretary of State, 1981), 883.

93. *Freedman's Bureau,* also known as the Bureau of Refugees, Freedmen and Abandoned Lands, was created by an act of Congress in 1865 to facilitate the relief and education of former slaves in their transition into free society. See Anderson, *The Education of Blacks.*

94. O. Bizzell, *The Heritage of Sampson County, North Carolina* (Winston-Salem, NC: Hunter Publishing Company, 1983).

95. Blu, *The Lumbee Problem;* Dial, *The Lumbee;* Dial and Eliades, *The Only Land I Know.*

96. Blu, *The Lumbee Problem;* Dial, *The Lumbee;* Dial and Eliades, *The Only Land I Know.*

97. Blu, *The Lumbee Problem;* Dial, *The Lumbee;* Dial and Eliades, *The Only Land I Know.*

98. An act to provide for separate schools for Croatan Indians in Robeson County, Laws of North Carolina, 1885, c. 51.

99. McPherson, "Indians of North Carolina."

100. Pierce, *The Lumbee Petition,* 53.
101. An act to amend the laws of 1885 and 1887 so as to provide additional educational facilities for the Croatan Indians; citizens of Robeson County, North Carolina, Laws of North Carolina, 1889, c. 60; An act to empower the trustees of the Indian Normal School of Robeson County to transfer title to property of said school by deed to State Board of Education, and to provide for the appointment of trustees for said school, Public Laws of North Carolina, 1911, c. 168.
102. An act to establish an Indian legislative committee for the Indian Schools of Robeson County, Public Laws of North Carolina, 1921, c. 146.
103. Blu, *The Lumbee Problem.*
104. Dial and Eliades, *The Only Land I Know.*
105. Ibid.
106. Sider, *Lumbee Indian Histories.*
107. Ibid.
108. Blu, *The Lumbee Problem;* Dial and Eliades, *The Only Land I Know;* Sider, *Lumbee Indian Histories.*
109. T. R. T. Massey, "School Desegregation: Its Significance for Native Americans of Eastern North Carolina" (master's thesis, University of North Carolina at Wilmington, 1996); V. R. Thompson, "A History of the Education of the Lumbee Indians of Robeson County, North Carolina from 1885 to 1970" (doctoral dissertation, University of Miami, 1973).
110. Fuchs and Havinghurst, *To Live on This Earth;* Lomawaima, "Educating Native Americans"; Spring, *Deculturalization;* Swisher and Tippeconnic, *Next Steps*
111. Cobb, *Listening to Our Grandmothers' Stories;* Ellis, *To Change Them Forever;* Fuchs and Havinghurst, *To Live on This Earth;* Lomawaima, *They Called It Prairie Light.*
112. "Kill the Indian to save the man" was a motto adopted by boarding schools to facilitate the deculturalization of American Indians. It is attributed to Captain Richard Henry Pratt, the founder of the Carlisle Indian Industrial School. See Richard H. Pratt, "The Advantages of Mingling Indians with Whites," in *Americanizing the American Indians,* ed. Francis Paul Prucha (Cambridge, MA: Harvard University Press, 1892), 260–261.
113. Bordewich, *Killing the White Man's Indian;* Cobb, *Listening to Our Grandmothers' Stories;* Ellis, *To Change Them Forever;* Lomawaima, *They Called It Prairie Light;* Lomawaima, "Educating Native Americans"; Marshall, *Cultural Diversity in Our Schools;* Swisher and Tippeconnic, *Next Steps;* Szasz, *Education and the American Indian;* Tatum, *"All the Black Kids."*
114. Bordewich, *Killing the White Man's Indian;* Cobb, *Listening to Our Grandmothers' Stories;* Ellis, *To Change Them Forever;* Lomawaima, *They Called It Prairie Light;* Lomawaima, "Educating Native Americans"; Marshall, *Cultural Diversity in Our Schools;* Swisher and Tippeconnic, *Next Steps;* Szasz, *Education and the American Indian;* Tatum, *"All the Black Kids."*
115. *Termination* refers to the period of federal Indian policy from the 1940s to the early 1960s. See Sokolow, *Native Americans and the Law.* The goal of termination was for Indians to become more self-sufficent and to assist in their assmiliation. See Fuchs and Havinghurst, *To Live on This Earth;* Vine Deloria, *American Indian Policy in the Twentieth Century,* 1st ed. (Norman: University of Oklahoma Press, 1985); Szasz, *Education and the American Indian.*
116. Fuchs and Havinghurst, *To Live on This Earth.*
117. *Allotment* refers to an assimilation policy wherein the tribal lands of the reservation were surveyed and divided and each individual member of the tribe was given a share of the land to farm. Allotment was in opposition to the tribes' tradition of

holding the lands jointly. This policy succeeded in breaking up the tribes' land and was also an effort to assimilate the Indians into white culture. See also Bordewich, *Killing the White Man's Indian*.

118. Szasz, *Education and the American Indian*.
119. V. Lambert, conversation with author, July 10, 2004.
120. Szasz, *Education and the American Indian*.
121. Fuchs and Havinghurst, *To Live on This Earth*.
122. John R. Finger, *Cherokee Americans: The Eastern Band of Cherokees in the Twentieth Century* (Lincoln: University of Nebraska Press, 1991); Finger, *The Eastern Band of Cherokees, 1819–1900* (Knoxville: University of Tennessee Press, 1984).
123. Finger, *Cherokee Americans; The Eastern Band of Cherokees*.
124. Sider, *Lumbee Indian Histories*.
125. Dial, *The Lumbee*.
126. Sider, *Lumbee Indian Histories*.
127. Ibid.
128. Ibid.
129. Pierce, *The Lumbee Petition*.
130. McMillan, *Sir Walter Raleigh's Lost Colony*.
131. The University of North Carolina at Pembroke started functioning in 1887 as an Indian Normal School for the training of teachers for the Lumbee community. In 1940 the school became recognized as an institution of higher learning and became Pembroke State College for Indians. Dial, *The Lumbee*.
132. Ibid.
133. Amendment to Croatan Indian Laws Act; Dial, *The Lumbee*.
134. Pierce, *The Lumbee Petition*, 40.
135. Ibid.
136. Ibid., 54.
137. Blu, *The Lumbee Problem*; McPherson, "Indians of North Carolina."
138. Blu, *The Lumbee Problem*.
139. Fuchs and Havinghurst, *To Live on This Earth*.
140. *An Act to Amend the Consolidated Statutes and to Codify the Laws Relating to Public Schools* (March 2, 1923), 311.
141. Sider, *Lumbee Indian Histories*.
142. An act to amend Section Five Thousand Four Hundred and Forty-Five of the consolidated statutes so as to provide keeping separate records for the public schools for the Cherokee Indians of Robeson County, State of North Carolina Public Laws and Resolutions, 1931, c. 141. In 1931, the Lumbee were referred to as Cherokee Indians of Robeson County.
143. Massey, "School Desegregation."
144. Ibid.
145. Ernest Dewey Hancock, "A Sociological Study of the Triracial Community in Robeson County, North Carolina" (master's thesis, University of North Carolina at Chapel Hill, 1935).
146. Ibid.
147. *Public Schools of Robeson County Board of Education (PSRCBE) Minutes*, (Lumberton, NC); Thompson, "A History of the Education of the Lumbee Indians of Robeson County, North Carolina from 1885 to 1970."
148. *PSRCBE Minutes*, Thompson, "A History of the Education of the Lumbee Indians."
149. *Brown v. Board* provided the hallmark supreme court decision against segregation by law, and also against the "separate but equal" doctrine. *Brown V. Board of Education of Topeka, Kansas*, 347 U.S. 483 (1954).

150. Massey, "School Desegregation."
151. Ibid.
152. *Jim Crow* refers to legal statutes passed during slavery, the Civil War, and the Reconstruction to segregate whites from nonwhites, to restrict the freedom of non-whites, and to perpetuate the dominance of whites and the subordinance of nonwhites; Woodward, *The Strange Career of Jim Crow.*
153. Massey, "School Desegregation."
154. Ibid.; Thompson, "A History of the Education of the Lumbee Indians of Robeson County, North Carolina from 1885 to 1970."
155. Patricia Barker Lerch, "State-Recognized Indians of North Carolina, Including a History of the Waccamaw-Sioux," in *Indians of the Southeastern United States in the Late 20th Century,* ed. J. Anthony Paredes (Tuscaloosa: University of Alabama Press, 1992).
156. Ibid.; Massey, "School Desegregation."
157. Ibid.
158. V. L. Jackson, "Asserting Identity: Coharie Indians and the East Carolina Indian School" (paper presented at the New Directions in American Indian research: A gathering of emerging scholars, Chapel Hill, NC, March 19, 2004).
159. Lerch, "State-Recognized Indians of North Carolina"; Massey, "School Desegregation."
160. Massey, "School Desegregation."
161. Ibid.
162. "North Carolina Indians: American Indian Education," newsobserver.com, June 25, 2004, modified April 10, 2006, http://www.newsobserver.com/ nie/curriculum/ social_studies/nc_indians/story/1368873p-7377987c.html (accessed October 20, 2005).
163. Thompson, "A History of the Education."
164. William S. Powell, *Higher Education in North Carolina* (Raleigh, NC: State Department of Archives and History, 1970).
165. Dial, *The Lumbee.*
166. McPherson, "Indians of North Carolina."
167. Ibid.; Pierce, *The Lumbee Petition.*
168. Powell, *Higher Education in North Carolina.*
169. Ibid.
170. J. L. Sanders, conversation with author, July 11, 2005; Sanders, "The University of North Carolina: The Legislative Evolution of Public Higher Education," *Popular Government* 59, no. 2 (1993).
171. William D. Snider, *Light on the Hill: A History of the University of North Carolina at Chapel Hill* (Chapel Hill: University of North Carolina Press, 1992).
172. Sanders, conversation with author, July 15, 2005.
173. An act repealing Articles 2 through 9 of Chapter 116 of the General Statues and substituting a new Article 2 in lieu thereof State of North Carolina Session Laws and Resolutions, 1957, c. 1142.
174. Blu, *The Lumbee Problem;* J. W. Currie, conversation with author, July 1, 2005; L. Oxendine, conversation with author, July 5, 2005.
175. Ibid; Nelson H. Harris, "Desegregation in North Carolina Institutions of Higher Learning," *The Journal of Negro Education* 27, no. 3 (1958); W.M., conversation with author, October 11, 2001; W.L., conversation with author, October 21, 2001; Richard, conversation with author, November 10 and 15, 2004; Currie, conversation with author, July 1, 2005.
176. Richard, conversation with author, November 10 and 15, 2004.

Conclusion: Reflections on the Historicality of Education Systems

Kim Tolley

What are the conditions under which education systems change? In seeking answers, social historians often look along trajectories, explaining the meaning of events by their relation to the unfolding of social experience. Whether the researcher takes a narrative view and constructs a story of a community or a state, or takes an analytic and quantitative approach, the focus is always the same: the sequential development of social outcomes. However, as the sociologist Andrew Abbot notes, "The social process doesn't have outcomes. It just keeps on going."[1]

This chapter synthesizes several concepts presented in the secondary literature and in this collection of essays. The following discussion begins with a brief overview of the theoretical work of Margaret S. Archer, Andy Green, and Ting-Hong Wong, and then turns to an analysis of the diverse case studies in this volume in light of several key concepts from the secondary literature.

The Historiography

Over the past several decades, a body of scholarship has developed to address the following questions: How do education systems develop? Why do they change over time? How can we explain differences in educational development among different countries, especially among those with similar colonial origins?

In what remains arguably the most comprehensive comparative study to date, the sociologist Margaret S. Archer explains educational transformations in terms of group interaction, including group conflict, the development of alliances, and the elaboration of effective ideologies. She seeks to account for transformations in the characteristics and processes of national education systems. Such characteristics and processes include problems of educational equity, access, the transmission of knowledge and values, social status, and social mobility. According to

Archer, dominant or controlling groups shape the characteristics of education systems: "[C]hange occurs because new educational goals are pursued by those who have power to modify previous practices." Her analysis of interaction in terms of group power, conflict, the promulgation of ideologies, and the formation of political alliances is similar to interpretations based on the concept of *hegemony.* The concept of hegemony, elaborated by Antonio Gramsci,[2] refers to the authoritative power of dominant groups in society to control social rhetoric, policy, infrastructure, and processes. What distinguishes Archer's approach is her theorization of how subordinate groups can challenge the status quo, and how such challenges can shape the subsequent form of educational systems within states.[3]

Archer theorizes that two forms of challenge to existing education systems are possible. The first is *substitution,* a process by which groups create rival institutions to obtain forms of schooling compatible with their interests and needs. The second is *restriction,* a political process by which groups destroy the monopoly of the dominant group through legal constraints and state provision. Archer argues that in all systems, education becomes increasingly integrated with the state and with a plurality of other local institutions, but this occurs in different ways. To explain variance in the degree of centralization or decentralization in education systems, she theorizes that those with restrictive origins tend to acquire strong unification and systemization, and weak differentiation and specialization. Systems with substitutive origins tend to develop with weak forms of unification and systematization and strong forms of specialization and differentiation. Thus, in all systems, education becomes increasingly integrated with the state, but this process evolves differently in centralized and decentralized systems.[4]

While the historian Andy Green acknowledges that Archer's study is the first to analyze the mechanisms of change in education systems, he faults her theory as "only half an explanation," because it does not situate an analysis of educational change within a comparative theory of the state. Green argues that the evolution of public education systems cannot be understood apart from the context of state formation. He uses Gramsci's theory of hegemony to explain the relations among the social forces involved in the exercise of state power. According to Green, *state formation* includes the building of political and administrative infrastructure, the development of a ruling ideology, and the consolidation of national consciousness. He argues that although economic relations might influence educational forms, it is the state that determines the shape of the education system, because competing demands on education are always resolved at the political level.[5]

Green concludes that the timing and development of education systems is shaped by the nature of the state and state formation. "[I]t was specifically the intervention of the state which effected the formation of national education systems, and it is therefore the nature of the state in different countries which must carry the largest burden of explanation for the particular national forms and periodizations of the development of school systems." According to Green, in all national contexts, the nature of class relations determines the forms and content of public education. He concludes that highly centralized states create centralized public education systems, whereas decentralized states develop decentralized systems. [6]

Both Green and Archer link the evolution of educational forms and processes to their origins. For Archer, the origin of such systems—whether these were substitutive or restrictive—determine whether they will develop highly centralized or decentralized forms. Green's conclusion that highly centralized states create centralized public education systems, whereas decentralized states develop decentralized systems, roughly echoes this premise. The difference between the two lies in the degree of power and control each allocates to the state. For Archer, the shape of education systems unfolds as a result of political struggles among powerful interest groups. From this perspective, there exists no "hidden hand," no central all-knowing authority with the ability to influence the characteristics of education systems. In contrast, although he acknowledges the role played by "particular configurations of class forces" in shaping major transformations in schooling, Green locates the central agency for change in the authority of the state.[7]

Recently, the social historian Ting-Hong Wong has challenged the notion that transformations in schooling occur as a direct result of state intervention. Wong argues that state formation is rarely the straightforward process portrayed by other scholars who have written on education and the state. Wong's comparative historical study of educational transformation in Hong Kong and Singapore leads him to conclude that the state can face multiple and contradictory pressures, a phenomenon neglected by previous theories of education and state formation. Wong defines state formation as "the historical process through which the ruling elites struggle to build a national or 'local' identity, outmaneuver political antagonists, and integrate the society." This process includes efforts to cultivate national identity, advance social integration, consolidate dominance, and win support from subordinate groups. As Wong points out, there is never any guarantee that a ruling regime can succeed in achieving all these goals; strategic compromises, concessions, and outright political failures often contribute to contradictory results.[8]

According to Wong, Green and other theorists have underestimated the relative autonomy of public school systems in some states and have failed to consider the extent to which a ruling group can establish its dominance by incorporating the culture of the subordinated groups. Because such acts of *cultural incorporation* ultimately influence the nature of the state itself, Wong theorizes that state formation and education are related in an interactive, dialectical, and recursive manner.[9]

This body of scholarship provides a number of key concepts from which to consider the essays in this volume. One of the benefits of the case-study approach is that it allows the researcher to test a broad theory against the historical development of social processes in a specific context. The following discussion begins by analyzing the diverse case studies in this volume in light of concepts from the secondary literature, such as the role of origins in influencing the evolution of centralized or decentralized systems of schooling, and restriction, substitution, and cultural incorporation as factors in educational transformations. Several authors in this volume identify additional factors in educational transformation, including the influence of international policy networks, and educational co-option by the state. This chapter discusses each of these factors and analyzes the role played by voluntary or market-based systems of schooling in each of these concepts.

The Role of Origins in Influencing the Evolution of Centralized or Decentralized Systems of Schooling

Both Archer and Green conclude that education systems evolve differently in centralized and decentralized systems. Archer locates this difference in the processes by which education systems originate within states. Green locates this difference in the national origins of the state and the subsequent process of state formation, and concludes that centralized states develop centralized education systems, whereas decentralized states develop more decentralized education systems.[10] However, several essays in this volume present findings that challenge these notions.

Kim Tolley and Nancy Beadie's chapter suggests that in the United States, education systems with different origins evolved to have very similar forms and processes. By the late twentieth century, both New York and North Carolina funded universal precollege schooling through a system of public schools. However, the origins of the education systems in both states could not have been more different. New York State had a relatively centralized organization in the early national period and provided at least some form of public funds to so-called common schools for the purpose of universal schooling. In contrast, North Carolina adopted a market-based approach to education, providing no public funds in support of schooling until 1840. This early system—the result of policy deliberations in the North Carolina state legislature—allowed various interest groups to create educational institutions compatible with their own religious ideology, culture, and class status.[11] Tolley and Beadie's chapter demonstrates that despite distinct differences in early national schooling in these two states, women's access to schooling underwent a similar transformation in both regions.[12]

Geoffrey Sherington and Craig Campbell found that although Australia developed a centralized system, the emergence of the concept of public education mirrored the American experience. By the late nineteenth century, an American public school could be described as an institution established and funded by the state to provide universal schooling to all citizens, and which was subject to inspection. According to Sherington and Campbell, this was also a reasonable description of public schools in Australia, although they note that whether all citizens could attain equitable schooling in any nation during the period is debatable. In both the United States and in Australia, the Protestant public school establishment sought to maintain its culture and religion in a society characterized by rapid demographic change as a result of immigration. Sherington and Campbell argue that in both countries, the trend toward the development of large, centralized systems of public schooling was associated with the emerging ascendancy of the Protestant middle class in both state administration and the politics of education.[13]

Tim Allender's chapter demonstrates that under colonial rulers, a national education system could originate with highly decentralized features and processes, evolve to a more centralized model, and shift back again. Early nineteenth-century British schooling in colonial India originated with the Orientalist approach, in which the state permitted its agents discretion to provide support to widespread indigenous forms of schooling. This policy approach shifted in 1854, when Charles Wood's Education Dispatch gave rise to a bureaucratic regulatory model of systemic state schooling in each province. However, state support for a decentralized educational system supportive of indigenous schooling ended after the Revolt of 1857. In response, Lord Stanley directed that only government schools should be relied upon to provide universal schooling, although each province was left to develop its own primary education. Allender shows that this more centralized model prevailed until 1871, when the colonial government adopted a new administrative strategy of decentralization. According to him, the incentive to decentralize the education system arose from a need to save money rather than from a desire to provide more effective or equitable forms of schooling. In the early twentieth century, the state transferred responsibility for education to the elected legislatures of each province and presidency. Allender argues that by 1919, large increases in population, famine, and inflation had undermined the goals of many of the state's education initiatives.[14]

National education in Colombia and Mexico transformed from centralized to decentralized systems. Meri Clark shows that in Colombia, the centralized system of education initially established by the new national government failed in the wake of insufficient funding and central control. As a compromise, the nineteenth-century state moved educational obligations to private hands, a shift that led to the emergence of private school associations in the 1830s. Such private education societies assumed responsibility for maintaining public schools and establishing private ones.[15] Victoria- Maria MacDonald and Mark Nilles point out that although Mexico's 1824 Constitution adopted Enlightenment ideals and required the provision of universal education, neither state nor federal funds were authorized for public schools. This left local communities to rely solely on their own resources. In the case of poor communities unable to afford the tuition charged by so-called private schools, the end result was a lack of educational institutions; in the case of wealthier communities, private secular or denominational schools developed in response to a demand for tuition-based schooling.[16]

These studies suggest that the origins of education systems do not always predict the subsequent shape and form of their evolution. Whether such systems have restrictive or substitutive origins or emerge in the context of centralized or decentralized states, any combination of factors can influence their future development, including such economic factors as the availability of financial resources or shifts in supply and demand in the education market; catastrophic events such as wars or famines; acts of substitution or restriction; changing balances of political power among ruling interest groups; acts of cultural incorporation and compromise on the part of the state to buttress its political legitimacy among its citizens; and policy shifts in response to political alliances and international trends in education.

Restriction, Substitution, and Cultural Incorporation as Factors in Educational Transformations

Although few essays specifically address the process of restriction, several case studies in this volume provide examples of the processes of substitution and cultural incorporation. From a top-down perspective, acts of substitution and cultural incorporation can shape forms of schooling in local communities; from a bottom-up perspective, they can also contribute to state formation by shaping national identity, policy, and practice. These processes interact in what Wong terms "the dialectical relation of education and state formation."[17]

Restriction and Substitution

Archer defines *restriction* as a political process by which groups destroy the monopoly of the dominant group through legal constraints and state provision.[18] The clearest example of restriction in this collection of essays is arguably the national education system established in Colombia after the revolution of independence. The revolution not only severed Colombians' ties with the Spanish Empire but also offered Colombians the opportunity to replace the pre-existing hierarchical social system with an egalitarian republic. Clark's chapter explores the problematic early attempts of Colombians to establish a national education system based on Enlightenment ideology.[19]

A number of the chapters in this volume support Archer's concept of *substitution*. Archer defined substitution as the process by which groups can challenge the hegemony of the state by creating rival, privately funded institutions.[20]

In several case studies in this book, social class, status, and race played a role in the motivation of a more powerful group to substitute its own forms of schooling for those of the state. For instance, Clark demonstrates that Colombian elites opposed public schools not only for religious reasons but also because of the cost. Race and social status appear to have played a role as well. Wealthy, white Colombians refused to pay taxes to support schools for the poor, nonwhite children in their districts. As a result, indigenous and black Colombians had far less access to schooling than their white peers.[21] Allender's study suggests that in India, the colonial state leadership promoted the process of substitution by pursuing a policy that encouraged local or market-based forms of schooling to evolve. In this case, the state's policy of decentralization allowed Western experimentation to flourish in urban centers, which benefited elites who enjoyed Raj patronage.[22]

Disenfranchised groups have also used substitution as a strategy in the face of restriction. According to MacDonald and Nilles, Spanish colonial policies in the Americas dictated the land, civil, and political rights of individuals in New

Spain according to their national origin, ethnicity, or race. The resulting social and political hierarchies lasted well into the twentieth century in the United States. As the former Mexican territories entered statehood, some racial groups lost the right to vote. For instance, when California attained statehood in 1849, it excluded blacks and Native Americans from citizenship, a disenfranchisement that extended to schooling, since the state's early school code stipulated that "Negroes, Mongolians, and Indians" be excluded from the state-supported public schools. In this political context, disenfranchised communities interested in schooling their youth had no choice but to create alternative forms of schooling.[23] Heather Kimberly Dial's chapter shows that in the socioeconomic and political structure of North Carolina, segregation and racial hierarchy supported a caste system in which whites represented the higher caste and blacks the lower caste. Whites maintained dominance through economic and political power and an ideology of inequality. When the North Carolina Constitutional Convention voted to disenfranchise free nonwhites, one result was the categorization of all Lumbee Indians as black and a denial of their Native American culture and heritage. In response to their exclusion from all-white schools, the Lumbee embraced segregation and established their own schools, both public and private.[24]

Cultural Incorporation

Wong argues that "a reciprocal and dynamic relationship exists between state hegemony and social movements." He characterizes this as a dialectical connection between education and state formation, a relationship that results from interconnections among the state, social movements, and the politics of cultural incorporation. Wong defines *cultural incorporation* as the process by which the state incorporates local institutions and accommodates the culture of subordinate authorities to strengthen its ruling position.[25]

Many essays in this volume provide examples of cultural incorporation in various contexts. For example, Wong's study of schooling in Singapore shows how British colonial authorities initially allowed Chinese schools to operate as private institutions funded and operated by the local Chinese community. When the majority of Chinese residents became enfranchised after the mid-1950s, the popularly elected governments bowed to political pressure and adopted a policy of supporting distinctly Chinese schools. Wong argues that by supporting a linguistically and culturally compartmentalized school system, the state compromised its goals of promoting cultural integration and a common national identity.[26]

Wong's theory of cultural incorporation may explain the development of national *diversity policies in* Canada. As Reva Joshee and Lauri Johnson point out in their chapter on Canadian educational policy, "Recognizing the rights of diverse groups was already part of the policy landscape by the time Canada officially became an independent nation-state in 1867." At first glance, Canada's early diversity policies are puzzling, particularly since they developed despite expressed national goals of assimilation. Although the Royal Proclamation of

1763 expressed the British aim of assimilating the French, assimilation did not occur. Instead, the Quebec Act of 1774 sanctioned the continued existence of a separate French legal system, language, and culture, and the Constitutional Act of 1791 created separate administrative units that led to the separation of British and French Canada. Wong's theory of cultural incorporation provides a possible explanation for this development. Joshee and Johnson note that Canada's diversity policies emerged during a time when assimilation was not demographically possible.[27] With a relatively large French-speaking population, cultural incorporation remained the pragmatic strategy of the nation's leading political groups.

Some essays in this volume suggest that the degree to which a state engages in acts of cultural incorporation depends on its motivation for schooling and on the race, social class, and political power of those it seeks to educate. As MacDonald and Nilles have shown, in Central and North America, formal education under Spanish rule occurred in the Catholic missions. Mission education was purposefully designed to replace Native American languages, religions, and culture with the Spanish language, Roman Catholic faith, and European customs. Citing an array of secondary sources, MacDonald and Nilles claim that there exists little evidence of cultural incorporation in this case. Although some priests may have learned native languages to communicate more effectively, the mission's role in the de-culturalization of Native Americans was extensive.[28]

When a specific group has political power, wealth, or status, the state's efforts to accommodate the group's culture can result in institutional or procedural changes to the state's education system and ethos. For instance, Hispanos in New Mexico repealed a law stipulating that control of tax-supported schools would lie with the territorial government, because they wished to maintain local funding control. According to MacDonald and Nilles, as an outcome of this ethnic group's political power, New Mexico, in contrast to other southwestern territories, embraced Spanish/English bilingual instruction in its public schools.[29]

The Influence of International Policy Networks

State educational policy is shaped not only by political and social structures and processes within the state, but also by the external political and economic context and policy environment. As Peter Kallaway points out, there is a lack of research on the relationship between international conferences and developments in colonial and mission education. His chapter explores the influence of the policy deliberations of major professional educational conferences on shifts in British colonial education between the First and Second World Wars. Kallaway concludes that in the context of global economic depression and the rise of totalitarianism in Germany, Italy, Japan, and the USSR, a clear shift in emphasis emerged at the conferences of the New Education Fellowship, marking a move away from the personal and individual development of the Progressive Era toward polices to promote economic growth and democracy. He tracks this shift through a review of international conference records.[30]

Kallaway also brings to light the sometimes problematic relationship between the international Christian missionary education network and the state. He demonstrates that the emergence of this network from the late nineteenth century was highly significant in shaping the development of educational debate, particularly during the interwar years. For its part, the missionary conference network focused on such issues as how to work within the political framework of colonialism and how to propagate Christianity while meeting the economic, social, medical, and educational needs of colonized peoples. Missionary leaders expressed the opinion that if the missions were to retain their influence in a changing international climate, missionary societies would have to cooperate with the colonial governments. However, during the interwar years, such cooperation grew increasingly problematic.[31]

According to Kallaway, the interwar debate over the relation of the missions and colonial governments heralded a transformation in the nature of missionary education. On the one hand, missionary reformers recognized the importance of a good working relationship with colonial governments as a means of gaining influence over state educational policy development. On the other hand, some reformers viewed such collaboration as problematic. Particularly in India, the West Indies, and British Africa, reformers argued that the missions should accommodate the needs of indigenous peoples in order to retain any influence at all in local communities. As a result, in some regions, the religious message of personal salvation became inescapably linked with the political message of national independence and freedom.[32]

Educational Co-option by the State

While the essays in this volume yield examples of restriction, substitution, cultural incorporation, and the influence of international policy networks, they also portray instances of government action that Kay Whitehead describes as *co-option*. Over time, the process of substitution can contribute to the form and shape of large national systems of schooling as the state appropriates or co-opts substitutive forms of schooling. Here, *educational co-option* is defined as the process by which the state appropriates preexisting voluntary or market-based educational structures or systems.

Whitehead's study of teaching families in nineteenth-century Australia demonstrates how the state co-opted preexisting forms of schooling. Prior to state intervention, the teaching family was a common feature in Australian schools. Under the 1851 Education Act, the teaching family was appropriated by the state in response to demand for sex-segregated schools. As Whitehead shows, state intervention privileged men over women as teachers. After the introduction of compulsory schooling in 1875, the state differentiated wage rates for men and women, justifying men's higher wages on the assumption that men would marry and support dependents while women would remain single, a policy that effectively marginalized married women. Whitehead concludes that for women, an important and unintended consequence of the individuation of wages was the facilitation of "economic and social conditions for single women teachers who were discursively positioned as 'new women' to unsettle patriarchal norms and contest the gender order."[33]

In several countries, early national systems of education appropriated—or co-opted—preexisting institutions and forms of schooling. The process of co-option transformed the nature of schooling both at the local and state levels. For example, by the time large systems of education spread across the United States, schoolrooms had opened to women, and women had gained a majority of schoolteacher positions. Nancy Beadie and I show how this process developed in response to market forces of supply and demand. Elsewhere, we have argued that one major historical legacy of private or market-based forms of schooling in the United States is the increased participation of women in education. When state legislators began to fund the expansion of state-supported systems of schooling later in the nineteenth century, the inclusion of women in most schools, as students or as teachers, was a given.[34] Large state-sponsored systems of education simply co-opted the social norms and institutional features that had developed in privately funded schools in earlier years.

In some cases, the state co-opted policies and practices that contributed to educational inequities. For instance, in North Carolina, when the state legislature began to disburse large amounts of funding in support of universal schooling after the Civil War, it supported a large system of preexisting segregated schools, including those of the Lumbee, a system that continued without successful legal challenge until the mid-twentieth century.[35]

The theories of substitution, cultural incorporation, policy networks, and co-option include a role for market-based or voluntary schools, because such schools—whether secular or religious—have often served as significant alternatives to state-supported schooling. Nevertheless, most theories of education and state formation keep the focus on the state as the central provider of schooling and overlook the historical significance of voluntary, market-based schooling.

The Significance of Voluntary, Market-Based Schooling

The most common portrayal of the emergence of national schooling remains the state-controlled expansion of publicly-funded schools. From this perspective, recent national movements to implement market-based forms of school funding and organization appear as novel reforms, which they are not. There exists a longstanding precedent for market-based schooling in the form of eighteenth- and nineteenth-century chartered academies, mission and church schools, and other entrepreneurial private institutions.[36]

A narrow definition of national education as a system directed and funded by the state presents problems of timing when considered from the perspective of early state formation. Quoting Michael Katz, Green defines national education systems as "carefully articulated, hierarchically structured groupings of schools, primarily free and often compulsory, administered by full-time experts, and progressively taught by specially trained staff."[37] Based on this definition, many states did not establish what might be termed "national education systems" until several generations after their origin; in states such as North Carolina or

Mississippi in the United States, there exists a more than one-hundred-year gap between the origin of the state and the establishment of an education system fitting Katz's definition. For Green, early national voluntary, market-based forms of schooling are significant only to the extent that they represent a "period of historical gestation"[38] that eventually gives rise to a formal system of schools funded by the state. However, several essays portray the provision of market-based schooling as a policy choice linked to the rhetoric of national identity.

During the early national periods of some states, the provision of voluntary, market-based schooling was at times a distinct policy choice. In Colombia, the government facilitated the growth of unregulated schools as a cost-saving measure in the face of scanty resources and inadequate funding; in Canada, India, and North Carolina, the state accommodated the desires of powerful interest groups interested in maintaining schools that would promote their own class values, cultural identity, and religious beliefs; in Singapore, the state strategically facilitated market-based schooling in an effort to accommodate subordinate groups.

In many late eighteenth- and nineteenth-century contexts, the government's decision on whether or not to fund schools reflected larger ideas of contemporary political economy that related to questions of state formation and national identity and purpose. Beadie argues that nineteenth-century politicians and philosophers in Great Britain and the United States agreed that public funds should be spent only on educational ventures that could not reasonably be supported by individual households. She cites the Scottish philosopher Adam Smith, who explained this principle in *The Wealth of Nations* (1776). Smith argued that the invisible hand of the market could stimulate production and consumption of various goods and services, including primary schooling. However, Smith allowed that higher forms of schooling required state support, because the cost of a university education was too great for an individual household to bear. From this perspective, colleges and universities required financial support from the state to survive, whereas voluntary forms of primary and intermediate schooling could be sustained through the market. Beadie's study of the first common school laws in New York (1812–1816) suggests that the pattern of state involvement in schooling began to change as demand for primary and intermediate schooling expanded among the general voting population. "Voting with their purses and their children's feet, [New Yorkers] shifted the balance of common schooling away from male heads of household, toward basic English instruction for children of both sexes." She concludes that local issues of school funding attained a larger political significance, since the question of school funding gave every family with school-age children a stake in political decision-making.[39] From this perspective, the expansion of market-based and voluntary forms of schooling prior to the development of large, state-funded systems of schooling is a significant component of the process of state formation.

As several chapters in this volume demonstrate, early national systems of state-supported schooling often co-opted preexisting educational institutions and processes developed in the context of voluntary and market-based schooling. For instance, in San Francisco, California, when the state initiated funds for public schooling, some preexisting denominational schools accepted funding, changed their names, and started to function as "public schools." In many such cases, the teachers, students, community, and school buildings remained the same; only the names on the buildings changed.[40]

As discussed earlier in this chapter, the state's co-option of voluntary and market-based forms of schooling has sometimes also included policies of decentralization. In the late eighteenth and early nineteenth centuries, the motivation of some states to allow voluntary schools to provide schooling arose from contemporary political ideology; in early national Colombia, allowing the market to provide schooling was a pragmatic response to a lack of financial resources. Today, policy makers in a number of countries advocate market-based school reforms, such as voucher initiatives, charter schools, or the use of public funds in support of so-called private schools.[41] In some cases, such reform efforts have a broad base of support because they appeal not only to those who believe that competition in the education market will improve schooling for the poor, but also to those interested in strengthening such alternative forms of schooling as church schools and other private institutions. Contemporary debates over market-based school reforms thus mirror the debates accompanying earlier periods of educational transformation.

The chapters in this volume suggest that social inequalities can persist in both highly centralized and decentralized systems. In highly centralized systems, groups with the greatest political power can prevail over others in establishing educational structures and processes that best meet their own class interests. In highly decentralized systems, subordinate groups may succeed in establishing alternative forms of schooling through acts of substitution, but such acts can have the unintended consequence of ultimately reinforcing the hegemony of more powerful groups. In all systems, power is always contested and recreated, but the outcomes of such interactions are far from predictable.

The Historicality of Education Systems

While the essays in this volume have focused on the contexts and causes of educational transformations, observers of schooling around the world might well conclude that in many cases, education systems change relatively slowly. For instance, although today more members of underrepresented groups have gained greater access to schooling in many countries, this shift has been gradual; at the highest levels of schooling, parity does not yet exist. Why are education systems resistant to change? The historicality of education systems provides one possible explanation.

Andrew Abbot argues that in contrast to corporations or formal organizations, social structures do not have much continuity over time. He defines social structures as neighborhoods, occupations, newspaper readerships, church congregations, social classes, ethnicities, technological communities, and consumption groups, often without formal records. Such structures can change quickly and easily, because their memories are widely distributed and their records often weak. Historicality, for Abbot, consists in biological, memorial, and recorded continuity.[42]

However, the artifacts left by such social structures contribute to the historicality of education systems. To the extent that social structures include social classes, consumption groups, and ethnicities, they are composed of the kinds of groups engaged in hegemonic struggles with the state over schooling. The outcomes of some of these struggles become recorded as procedures, policies, or legislative acts. Such records are historical artifacts of past negotiations, compromises, and concessions. Through these records, large amounts of the past are brought into the present, as given educational practices, accepted pedagogical ideologies, and school laws. The mass of this substantive historical experience, preserved as rules, procedures, laws, and habits, shapes the social context of the education system, and it constitutes a given set of possibilities and constraints within which each generation must work in the present.

Most of the factors that contribute to educational transformation work slowly, over successive generations. In some cases, transformation occurs so gradually that education systems can appear fixed; nevertheless, as the authors in this volume have demonstrated, even states with highly centralized origins can develop more decentralized education systems over time, and vice versa. Although there are big historical structures that somehow span long periods, producing what Abbot refers to as an illusion of long, enduring historicality, for certain kinds of social structures, transformation is possible—even for big structures such as education systems.

Notes

1. Andrew Abbot, "The Historicality of Individuals," *Social Science History* 29 (Spring 2005): 1–13, 2.

2. See Antonio Gramsci, *Selections from the Prison Notebooks* (London: Lawrence & Wishart, 1971); Chantal Mouffe, "Hegemony and Ideology in Gramsci," in *Gramsci and Marxist Theory*, ed. Chantal Mouffe (London: Routledge and Kegal Paul, 1979), 179–181.

3. Margaret S. Archer, *Social Origins of Educational Systems* (London: Sage Publications, 1979), 2.

4. Archer, *Social Origins of Educational Systems*.

5. Andy Green, *Education and State Formation: The Rise of Education Systems in England, France and the USA* (New York: St. Martin's Press, 1990), 74.

6. Ibid, 309–310.

7. Ibid, 110.

8. Ting-Hong Wong, *Hegemonies Compared: State Formation and Chinese School Politics in Postwar Singapore and Hong Kong* (New York: Routledge, 2002), 26.

9. Wong, *Hegemonies Compared*.

10. Archer, *Social Origins*.

11. For a case study of market-based schooling in North Carolina, see Kim Tolley, "A Chartered School in a Free Market: The Case of Raleigh Academy, 1801–1827," *Teachers College Record*, 107, 1 (2005): 59–88.

12. Kim Tolley and Nancy Beadie, "Socioeconomic Incentives to Teach in New York and North Carolina," in this volume; on the history of academies, see Nancy Beadie and

Kim Tolley, eds, *Chartered Schools: Two Hundred Years of Independent Academies in the United States, 1727–1925* (New York: Routledge, 2002).

13. Geoffrey Sherington and Craig Campbell, "Middle Class Formations and the Emergence of National Schooling: A Historiographical Review of the Australian Debate," in this volume.

14. Tim Allender, "How the State Made and Unmade Education in the Raj, 1800–1919," in this volume.

15. Meri Clark, "Disciplining Liberty: Early National Colombian School Struggles, 1820–1840," in this volume.

16. Victoria-María MacDonald and Mark R. Nilles, "From Spaniard to Mexican and Then American: Perspectives on the Southwestern Latino School Experience, 1800–1880," in this volume.

17. Wong, "Education and State Formation Reconsidered: Chinese School Identity in Postwar Singapore," in this volume.

18. Archer, *Social Origins*.

19. Clark, "Disciplining Liberty," in this volume.

20. Archer, *Social Origins*.

21. Clark, "Disciplining Liberty," in this volume.

22. Allender, "How the State Made and Unmade Education in the Raj," in this volume.

23. MacDonald and Nilles, "From Spaniard to Mexican," in this volume. For discussion of academies and venture schools as alternative forms of schooling for disenfranchised groups, see Beadie and Tolley, *Chartered Schools*.

24. Heather Kimberly Dial, "Struggling for Voice in a Black and White World: Lumbee Indians' Segregated Educational Experience in North Carolina," in this volume.

25. Wong, *Hegemonies Compared*, 30.

26. Wong, "Education and State Formation Reconsidered," in this volume.

27. Reva Joshee and Lauri Johnson, "Historic Diversity and Equity Policies in Canada," in this volume.

28. MacDonald and Nilles, "From Spaniard to Mexican," in this volume.

29. MacDonald and Nilles, "From Spaniard to Mexican," in this volume.

30. Peter Kallaway, "Conference Litmus: The Development of a Conference and Policy Culture in the Inter-War Period with Special Reference to the New Education Fellowship and British Colonial Education in the Southern Africa," in this volume.

31. Ibid.

32. Ibid.

33. Kay Whitehead, "The Teaching Family, The State, and New Women in South Australian Schooling," in this volume.

34. Beadie makes this point in Beadie and Tolley, "Legacies of the Academy," *Chartered Schools*, 333.

35. Dial, "Struggling for Voice," in this volume.

36. Beadie and I have elaborated this point *in other work:* See Tolley, "A Chartered School in a Free Market"; Beadie "Toward a History of Education Markets in the United States: An Introduction" (paper presented at the Social Science History Association Annual Meeting, Chicago, 2004).

37. Andy Green, *Education and State Formation*, 2.

38. Ibid., 1.

39. Beadie, "Toward a History of Education Markets," 18.

40. Kim Tolley, "Consumerism, Education Markets, and Schooling in Early National California and North Carolina" (paper presented at History of Education Society Annual Meeting, Baltimore, MD, 2005).

41. For an overview of such debates in Great Britain and the United States, see Carnegie Foundation for the Advancement of Teaching, *School Choice* (Princeton, NJ: Carnegie Foundation, 1992); J. E. Chubb and T. M. Moe, *Politics, Markets, and America's Schools* (Washington DC: Brookings Institution, 1990); Peter W. Cookson, ed., *The Choice Controversy* (Newbury Park, CA: Corwin Press, 1992); Michael Engel, *The Struggle for Control of Public Education: Market Ideology vs. Democratic Values* (Philadelphia, PA: Temple University Press, 2000); B. Fuller, ed., *Inside Charter Schools: The Paradox of Radical Decentralization* (Cambridge, MA: Harvard University Press, 2002); Gerald Grace, *Catholic Schools: Missions, Markets and Morality* (New York: Routledge, 2002).

42. Abbot, "The Historicality of Individuals."

Contributors

Tim Allender is lecturer in the history of education/history curriculum at the University of Sydney. He is coauthor of *Coaches Called Here* (Adelaide: Advertiser Publishing, 1996), a social history of Western Victoria, Australia, which includes a survey of community-schooling traditions. More recently, he has published several articles in leading international journals on community, mission, and state-schooling enterprises in colonial India. He is currently South, West, and Central Asia editor of the *Asian Studies Review.* Following his research in Great Britain, Pakistan, and India, he published *Ruling through Education: The Politics of Schooling in the Colonial Punjab* (New Delhi: New Dawn Press, 2006). This study details how government and mission education evolved in response to the Raj in one large province in mid-nineteenth century.

Nancy Beadie is historian of education and associate professor in the area of educational leadership and policy studies at the University of Washington, Seattle. Her research focuses on nineteenth-century school politics and finance, especially as they shaped academies; the history of women's education; and school-community relations. She is coeditor, with Kim Tolley, of *Chartered Schools: Two Hundred Years of Independent Academies in the United States* (New York: Routledge, 2002). She has published articles in the *History of Education Quarterly, American Journal of Education, Teachers College Record, History of Higher Education Annual,* and *Education Policy.* In 2000, she received the History of Education Society's prize for best article in a refereed journal. She is currently writing a narrative history of one rural community and its schools in nineteenth-century New York.

Craig Campbell is senior lecturer in the history of education at the University of Sydney; author of one of the very few academic studies of an Australian high school, *State High School: Unley, 1910–1985* (Adelaide: Published by the author, 1985); and the major author of *Toward the State High School in Australia: Social Histories of State Secondary Schooling in Victoria, Tasmania and South Australia, 1850–1925* (Sydney: Australian & New Zealand History of Education Society, 1999). He has published many articles of significance, including "The State

High School in History," *Change* 4, no. 1 (2001): 5–18. With Geoffrey Sherington, he has edited *Education and Ethnicity* (2001) and has authored a review of the discipline, "The History of Education: The Possibility of Survival," *Change* 5 no. 1 (2002): 46–64. The results of their research collaboration were published in early 2005 as *The Comprehensive Public High School: A Study of Public Policy, Private Choice and Regional Diversity* (New York: Palgrave Macmillan).

Meri L. Clark is assistant professor at Western New England College. She teaches Latin American history, and her research focuses on the intellectual and cultural history of early republican Colombia. She received her BA from Reed College in 1995 and her PhD from Princeton University in 2003. She is now revising her dissertation, *Education for a Moral Republic: Schools, Reform, and Conflict in Colombia, 1780–1845*, for publication.

Heather Kimberly Dial is assistant professor in reading education at the University of North Carolina, Pembroke. She received her doctorate in the Curriculum Studies Program from North Carolina State University in 2005. She is a fully enrolled member of the Lumbee tribe. She has presented her research concerning the segregated Indian schools of the Lumbee at regional and national conferences, including the ERA and the National Indian Education Association Convention. Her research concerns are the segregated American Indian schools of the past, Lumbee Indians, culturally relevant teaching/curriculum, preschool education, and reading education. She is actively involved as an educator, supporter, and contributor to the positive development of her fellow Native Americans.

Lauri Johnson is associate professor of urban education in the Department of Educational Leadership and Policy at the University at Buffalo, New York. A former administrator with the New York City Public Schools, she completed a PhD in multicultural education at the University of Washington in 1999. Her research interests include examining how white educators conceptualize race, the historical development of multicultural policy and curriculum in the United States and Canada, and community activism in urban school reform. She has coedited *Urban Education with an Attitude* (Albany: SUNY Press, 2005) with Mary Finn and Rebecca Lewis. Her most recent book, coauthored with Reva Joshee, is *Multicultural Education Policies in Canada and the United States* (Vancouver: University of British Columbia Press, 2007).

Reva Joshee is assistant professor in the Department of Theory and Policy Studies in Education, Ontario Institute for Studies in Education, University of Toronto. Her focus is on the relationship between theory, policy, and practice in diversity education. She is currently involved in two major research initiatives: one focusing on the use of participatory policy analysis in the development of social justice policy in Canada and the other comparing diversity in education in India, Kenya, and Canada. Her published works include

"Citizenship and Multicultural Education in Canada: From Assimilation to Social Cohesion," in *Diversity and Citizenship Education: Global Perspectives,* ed. J. A. Banks (San Francisco: Jossey-Bass, 2004), 127–156; "A Framework for Understanding Diversity in Education in India," *Race, Ethnicity and Education* 6, no. 3: 283–297; "Canadian Multiculturalism as a Creature of Contraction and Expansion," *New Scholars-New Visions in Canadian Studies,* Fall 1996; and "The Federal Government and *Citizenship Education for Newcomers,*" *Canadian and International Education,* December 1996. Her most recent book, coauthored with Lauri Johnson, is *Multicultural Education Policies in Canada and the United States* (Vancouver: University of British Columbia Press, 2007).

Peter Kallaway is professor of comparative education at the University of the Western Cape, South Africa. He is coeditor (with G. Kruss, A. Fataar, and G. Donn) of *Education after Apartheid: South African Education in Transition* (Cape Town: UCT Press, 1997). He has published chapters and articles on South African education and the history of rural and industrial education in South Africa. He is author of "Forging a Nation," a video history of South African education and a study of key issues of contemporary policy, with One World Media, Johannesburg. He has served as a member of editorial boards of such journals as *Pedagogica Historica* (Ghent, Belgium), *History of Education* (UK), and *Southern African Review of Education.*

Victoria-María MacDonald is visiting associate professor, Department of Education Policy and Leadership, University of Maryland, College Park. She is editor of *Latino Education in U.S. History* (New York: Palgrave Macmillan, 2004). She has published articles in the *History of Education Quarterly* and *Educational Researcher.* Her regional specialty is the American South. Her fields of study include the history of education, the social history of American teachers, higher education, and Hispanic education.

Mark Nilles holds a bachelor's degree in journalism from the University of Wisconsin at Madison and a master's degree in international/intercultural development education from Florida State University. He has traveled throughout North America, Europe, West Africa, and Southeast Asia, and lived in Mali for two years as a Peace Corps volunteer, supporting small business and entrepreneurial activities. He currently works to improve education and employment opportunities for young people at the International Youth Foundation in Baltimore. He also serves as a board member for the Friends of Mali and the Youth Activism Project, Inc.

Geoffrey Sherington is professor (history of education) in the Faculty of Education and Social Work at the University of Sydney. He is an author and coauthor of an impressive list of books that includes *Fairbridge: Empire and Child Migration* (London: Woburn Press, 1998), *Youth in Australia: Policy Administration and Politics: A History since World War II* (Melbourne: Macmillan, 1995), and *Learning to Lead: A History of Girls' and Boys' Corporate Secondary*

Schools in Australia (Sydney: Allen and Unwin, 1987). With Craig Campbell, he has edited *Education and Ethnicity* (2001) and has written a review of their discipline, "The History of Education: The Possibility of Survival," *Change* 5, no. 1 (2002): 46–64. The results of their current research collaboration was published in early 2005 as *The Comprehensive Public High School: A Study of Public Policy, Private Choice and Regional Diversity* (New York: Palgrave Macmillan).

Kim Tolley is associate professor in the School of Education and Leadership at Notre Dame de Namur University. She is the author of *The Science Education of American Girls: A Historical Perspective* (New York: Routledge, 2003), and also coeditor (with Nancy Beadie) of *Chartered Schools: Two Hundred Years of Independent Academies in the United States, 1727 – 1925* (New York: Routledge, 2002). Her previous publications have appeared in the *History of Education Quarterly, Teachers College Record,* and the *Journal of Curriculum Studies.* Her research interests include the sociology and culture of teaching and learning in the late eighteenth and early nineteenth centuries, and the history of science and schooling. With the support of an Archie K. Davis research fellowship from the North Caroliniana Society, she is currently writing a book about the experiences of the American antebellum teacher Susan Nye Hutchison.

Kay Whitehead is associate professor in the School of Education at Flinders University, Australia, where she teaches the history and sociology of education. Her research has focused on teachers during the period from the mid-nineteenth to the mid-twentieth century. More recently she has expanded her work to include all women state employees in the postsuffrage era. She has published widely in history of education and feminist history journals, and is the author of *The New Women Teachers Come Along: Transforming Teaching in the Nineteenth Century* (Sydney: Australian & New Zealand History of Education Society, 2003).

Ting-Hong Wong received his PhD from the University of Wisconsin-Madison in 1999. He is an assistant research fellow at the Institute of Sociology, Academia Sinica, Taiwan. His first book, *Hegemonies Compared: State Formation and Chinese School Politics in Postwar Singapore and Hong Kong* (New York: RoutledgeFalmer) appeared in 2002. This book won the Award for Outstanding Publication by Young Scholars from the Academia Sinica in Taiwan in 2002 and received very positive reviews by leading education journals such as *Anthropology and Education Quarterly, Comparative Education Review, History of Education Quarterly,* and *International Studies of Sociology of Education.* Wong also has articles published in *British Journal of Sociology of Education, Comparative Education Review,* and *Journal of Historical Sociology.* He is now developing a new research project comparing education politics in postcolonial Taiwan and Singapore.

Index

Printed in the United States
By Bookmasters